T0226426

THE BIOLOGY OF THOUGHT

THE BIOLOGY OF THOUGHT

A NEURONAL MECHANISM IN THE GENERATION OF THOUGHT —A NEW MOLECULAR MODEL

KRISHNAGOPAL DHARANI

AMSTERDAM • BOSTON • HEIDELBERG • LONDON
NEW YORK • OXFORD • PARIS • SAN DIEGO
SAN FRANCISCO • SINGAPORE • SYDNEY • TOKYO

Academic Press is an imprint of Elsevier

Academic Press is an imprint of Elsevier
32 Jamestown Road, London NW1 7BY, UK
525 B Street, Suite 1800, San Diego, CA 92101-4495, USA
225 Wyman Street, Waltham, MA 02451, USA
The Boulevard, Langford Lane, Kidlington, Oxford OX5 1GB, UK

Notices
Knowledge and best practice in this field are constantly changing. As new research and experience broaden our understanding, changes in research methods, professional practices, or medical treatment may become necessary.

Practitioners and researchers must always rely on their own experience and knowledge in evaluating and using any information, methods, compounds, or experiments described herein. In using such information or methods they should be mindful of their own safety and the safety of others, including parties for whom they have a professional responsibility.

To the fullest extent of the law, neither the Publisher nor the authors, contributors, or editors, assume any liability for any injury and/or damage to persons or property as a matter of products liability, negligence or otherwise, or from any use or operation of any methods, products, instructions, or ideas contained in the material herein.

ISBN: 978-0-12-800900-0

Library of Congress Cataloging-in-Publication Data
A catalog record for this book is available from the Library of Congress

British Library Cataloguing-in-Publication Data
A catalogue record for this book is available from the British Library

For information on all Academic Press publications visit our website at http://store.elsevier.com

Typeset by MPS Limited, Chennai, India
www.adi-mps.com

Contents

Preface

When I began my work on this book a few years ago, I told my wife, with misplaced confidence, that it would be ready for publication within a year – but it has taken nearly five nerve-wrenching years for the book to be ready for human consumption! When I initially formulated the concept of the Molecular-Grid Model of Thought, I was under the erroneous impression that my model would be understood (and even that it would be readily welcomed!) by the readers, just like that – as if I were a conjurer who could pull rabbits out of a hat! But when I had shown the initial stages of my work to a few of my learned friends for comment, all I received were empty stares – and I soon realized that my presentation was half-baked and unfit for consumption, as it were. This experience was invaluable to me, for this has led to an elaborate discussion accompanied by many corrections, and this has eventually led to the development of this consummate work.

My first remedial task was to convince the reader that there is *no existing theory* in modern science that would explain this (seemingly) simple phenomenon of thought production in the brain. I asked my colleagues to verify the truth, and these modern times are so convenient in the sense that *anything* can be verified in a jiffy – you only need to Google it. Having firmly established that no such theory existed, my next task was to answer a flat statement from most readers: 'Is not the *whole* neuron responsible for generation of thought' – and this prompted me to dedicate an entire chapter (Chapter 6) to dispel this conviction! Now the reader is ready to take a bite into the book, and will soon be warming up to the logic and simple analysis presented in the chapters. The new model is convincing to the reader because it complies with modern concepts in neurobiology – and, adequately enough, it also explains the molecular basis of various mysterious phenomena of the brain, such as memory, forgetfulness, and intelligence. So I was happy and confident with the model and thought my job was almost over.

But soon I found myself stuck with another impediment, from another quarter of my friends, who came up with questions such as: 'Does it mean that the author has defined *what* exactly is thought – *simply* based on some electrochemical events occurring at the neurons?'; 'And was the ancient mystery of thought dispelled overnight?' – 'Not at all!' was my reply. I had to confide that I have kept the secret of thought intact in the good old antique vault, but that I have only attempted, in this book, to show where (and how) exactly does a thought originate in the neurons; and the true nature of thought is as much an enigma as ever. And this discussion led to the development of Chapters 9 and 10, which explain some ambiguous phenomena of the brain, such as the mind and consciousness, and the vexing mind-and-matter problem.

Finally, my friends and well-wishers complained that all these things are alright with neuroscientists and the like, but what of the science readers who are not nerve specialists, and what of the philosophers, in whose hands the book is likely to land? How will they understand these ideas? I realized that all these people certainly deserve some introduction to the basics of neurobiology in order to get a good grasp of the model. Now I had to get down to brass-tacks, and the outcome is the first three large chapters dealing with the fundamentals of the brain.

Having said these things, I should admit that work of this sort is formidable to write and is akin to navigating in uncharted seas – you must take the established facts of science and regard them in a new light, and work them up to explain newer concepts. And, as you will see shortly, I have succeeded in doing this. I have every reason to believe that this work paves the way to clear many of the abstract mysteries shrouding the field of neuroscience.

I wish the reader all the best, and a pleasant journey through this book!

Krishnagopal Dharani

Introduction

WHAT IS 'THE BIOLOGY OF THOUGHT'?

The human brain is the thinking organ, and neurons are its fundamental units. The myriad thoughts emanating from the magnificent human brain, minute after minute, are the products of the functioning neurons – and therefore we can say that neurons are the thinking machinery of the brain. However, it is still a puzzle how these neurons are designed to produce cascades of thoughts and can store them up in seemingly endless memory – all we know is that the sensations we receive from the outside world leave some sort of molecular signatures inside our brains that are imprinted in some form for future retrieval. What are these molecular signatures, and how are they made? How are these memory traces stored in the neurons? Our present science has no comprehensive answer – not yet.

The question of 'what is thought' is almost as ancient as human civilization. Thought is an abstract phenomenon occurring in the human brain. Thoughts underlie all human actions and interactions, and they make up the core of human *mind* and the background of human consciousness. The idea of 'thought' may mean so many different things to each of us. We have thoughts of material things such as flowers and mountains; or we may think of such abstract things such as the beauty of nature, solution to a problem, principle of a theory, etc.; or we have thoughts related to the past or future, and so on. These mental processes are so varied and diverse that it is impossible to group all these thoughts into a single definition. And all these thoughts are complex cerebral events requiring several areas of the brain operating together to produce coherent ideas. However, we are not concerned with such complex thoughts here in this work.

'Thought', in this book, means the most fundamental unit of mental processes happening in the brain. This work is concerned with the generation of the most basic unit of thought right at the level of neurons. There is no current theory in modern neuroscience as to how a stimulus from the external world is converted into a perception in the brain. Clearly, the human brain has some mechanism for this conversion. For example, we can perceive the color blue only when we are capable of converting the color blue we receive from the external world into internal thoughts of blue color in the brain. On the other hand, we fail to perceive ultraviolet light (or infrared rays, for that matter) because we do not possess the necessary mechanism to convert the UV light wavelengths into internal signals (as bats and butterflies do). Where is this mechanism located in the neurons? And how does this mechanism function? These questions are left unexplored and unanswered by modern science, and thus have remained a mystery until now.

This book, *The Biology of Thought*, is about the mechanism by which thought is generated and stored in the human brain. It describes a new molecular model by which thought is generated at the micro level in the human brain – right at the level of neurons. This work demonstrates how the biological electrochemical events occurring at the neuron interact with its molecular mechanisms to generate thoughts. In other words, here we have laid *biological foundations* for the generation of thought, hence the title of this book, *The Biology of Thought*. Thus, the hitherto *abstract thought* is shown to have a solid *physical origin* in the neuron, and this is showcased in a hypothesis called the *'Molecular-grid Model'*.

HOW IS THIS BOOK ORGANIZED?

Now we can see that to study such a complex issue as the generation of thought in the brain, the reader must have an overall idea of the anatomical and functional aspects of the brain to assist him/her to fully understand the mechanism. For easy sailing, the book is presented in four parts.

Part I (***Basic Concepts in Neurobiology***). Part I of the book consists of three chapters which gives us some basic information on the structure of the brain and its functioning. Chapter 1 deals with the general plan of the human nervous system and the anatomy of the brain. Chapter 2 deals with the functioning of neurons – all the complicated electrochemical events occurring in a neuron are discussed in a simple and lucid way so that even a non-physiologist can grasp the electrical properties of neurons. Chapter 3 discusses the phenomenon of memory – the current concepts of the formation of memory along with the molecular basis of its storage are described briefly. Part I is obviously meant for those readers who want to brush up their knowledge on these basic concepts of the brain; however, this discussion may also interest a neuro-academic, because the presentation is a little out of the ordinary, to suit the purpose of the book.

Part II (***The Molecular-Grid Model***). This part of the book deals with the crux of the problem of the generation of thought by neurons. It consists of five chapters which attempt to answer the following three questions:

1. What is the fundamental unit of thought?
2. Where is it generated in the neurons?
3. How is it generated?

The sensory receptors sense various stimuli from the external world and send them up to the neurons in the brain, where they are converted into perceptions. This initial step generates the most fundamental units of thought in the neurons. Chapter 4 examines these fundamental units of thought (called the 'primary thoughts'), and describes the properties of these primary thoughts. In Chapter 5 it is shown conclusively that only certain types of sensory neurons are designed to convert the external stimuli into primary thoughts (thus introducing the concept of 'perceptual neurons'). In Chapter 6 it is shown that neurons have a certain molecular mechanism to receive the external stimuli and convert them into primary thoughts. A scientific discussion is undertaken to study the precise location of these molecular gadgets in the neurons. Chapter 7 examines the structure of these molecular gadgets (now called the molecular grids) and shows how they operate to generate primary thoughts. Chapters 4–7 take all the well-established concepts in neuroscience and show them in a new light, which helps us reach some far-reaching conclusions – and this can finally lead us to the concept of the molecular-grid model.

Having proposed a working model for the generation of thought by the neurons, we should now see how it works – now the molecular-grid model has to pass the acid test. As with any good scientific model, it must be able to explain at least a majority (if not all) of the concerned phenomena, and thus our new model must now explain the most obvious phenomenon of the brain, that is the phenomenon of *memory*. Therefore, Chapter 8 applies the molecular-grid model to explain the molecular basis of memory and its storage – and alongside, also explains other equally intriguing features of the brain, such as forgetfulness, intelligence, and such others. Part II is the work-book of a theoretical scientist – because it consists of all the new concepts in the field of neuroscience, which can open new vistas in the field of neuroscience.

Throughout Part II, thought is dealt with as an *electrochemical process* – i.e., a stimulus traveling to the neurons, exciting them electrochemically and generating primary thoughts. But it should be pointed out to the reader that the phenomenon of thought is far more elusive. We have seen, in the preceding chapters, only a physical mechanism by which thoughts are generated, but we have studied nothing about mind and consciousness or about the exact nature of thought itself.

Part III (***The Evolution of Human Mind***). The human mind is not a single entity but has many varied aspects – such as perceptions, cognition, memory, reasoning, emotions, desires, etc. Likewise, consciousness is also a vague phenomenon. Historically, mind and consciousness have eluded a precise definition, mostly because they cannot be examined by a strict scientific method. Part III consists of two chapters. Chapter 9 engages the reader in a scientific discussion of mind and consciousness, and arrives at important definitions. Alongside, we also examine intriguing aspects of the human mind, like intelligence and emotions.

Chapter 10 treats *thought as a metaphysical entity* (in contrast to the electrochemical process described in Part II). The ambiguous issues dealt with here are very complex; however, we will deal with them in a stepwise fashion to unravel the mystery of thought, and we approach these philosophical aspects of thought in a strictly scientific manner (or, at least, as strictly as abstract philosophical thought allows). At the end, we will examine the perplexing metaphysical conundrum that has bothered philosophers for ages – the mind-and-matter problem – and see how the molecular grids become relevant in this discussion.

The philosophical presentation in Chapter 10 may appear a little out-of-place to a student of biology, but even a pragmatic biologist who studies human thought dispassionately must be aware of these hard questions to remind him/her of the limitations of human knowledge. And biologists must also realize that there are a lot of loose ends in the fabric of our knowledge in the field of neuroscience, which young scientists will have to unravel in future.

Part IV (***The Computer and the Brain***). It is customary to compare the brain to a computer – but here in Chapter 11 we will compare specifically the molecular grids in the neuron with transistors in a computer, and find striking similarities between them.

AUTHORS' NOTE: ON THE USES OF THE MOLECULAR-GRID MODEL

Scientific progress can be achieved chiefly in two ways – one, by way of observation and painstaking laboratory experimentation; and the other by way of conceptualization. Conceptualization is the formation of an abstract principle in the mind of a researcher in order to answer the question under observation, basing it upon the available evidence. Experiments and concepts are complementary to each other – a concept leads to experimental verification, and experimentation proves a concept correct. However, history of science shows that new concepts have radically changed the direction of experimental science and, consequently, the future of mankind itself. Some of the great discoveries in the field of science were possible with concepts being propounded first, which were proved correct by experiments much later; here are two examples in biology. Consider how DNA's helical structure was discovered. For a long time scientists had wondered why the four bases in DNA occur in exact proportions, but James Watson and Francis Crick (1953) made a conceptual breakthrough by proposing that the paired bases were arranged in repeated intervals in a double-helix pattern – and this model had rightly answered many bewildering questions in genetics and thus opened new horizons in genetic engineering. Another example is that of oxidative phosphorylation in the mitochondria (which is the last stage in cellular respiration and the final step in the generation of ATP in a cell). When the whole world was fruitlessly searching for some sort of chemical intermediates in the mitochondria that could transfer energy to ATP, it was the British biochemist, Peter Mitchell (1961), who proposed his ingenious model called chemiosmotic coupling, which showed how the trick lies in the transfer of energy by electrochemical gradient to generate ATP, and this concept has revolutionized the field of cell biology.

We will now ask, coming back to our present concern: 'Of what use is this molecular-grid model?' The outstanding achievement of this work, as we can see, is the precise molecular delineation of thought generation in the neurons – all for the first time in the history of human thought! In addition, this model has offered a full explanation for the formation and storage of memory – with unprecedented clarity as to what goes on in the brain during memory formation, much beyond the synapses and neurotransmitters. Alongside, much to its credit, this model has explained the workings of the most intriguing features of the human brain, such as intelligence, forgetfulness, mind, consciousness, etc. Apart from whatever has already been explained, the new model has the promising potential to cause a conceptual change in further research in neuroscience – giving the experimental neuroscience a new impetus. There are many voids in neuroscience; for example, the pathogenesis of most mental afflictions is far from clear, and consequently their treatment is by and large unsatisfactory. The important reason for these impediments in progress, one can presume, is the lack of a proper concept of the workings of the brain. Moreover, present-day investigations in brain research, such as electroencephalogram (EEG), magneto-encephalogram (MEG), positron emission tomography (PET) and functional magnetic resonance (*f*MRI), are but *indirect* measurements of brain activity and thus are inadequate for a proper study of the brain; whereas some direct measurement techniques, such as electrode implantation in the brain, are not as acceptable ethically in humans as perhaps they are in animals. This model surely expands the frontiers of human knowledge in these fields.

This book also has an additional, yet important, outcome. The cognitive issues undertaken in Chapters 9 and 10 have led to some significant assertions hitherto unasserted – surely, many cognitive scientists and philosophers may find these conclusions not only interesting but invigorating as well. However, it is justified that some scholastic philosophers, after reading Part III, would curl their lips in distaste – the complaint surely to be over the profound clarity of ideas presented in there and their sheer simplicity – nothing is more damaging to a philosopher than being clear.

Finally, it must be admitted that the new concepts described in this work may run counter to the established notions we have cherished all along – but the strong reasoning offered in this work will eventually lead the reader to convince himself/herself of these concepts. However, when we have agreed with these new concepts, we may find ourselves disagreeing with our teachers in the older theories of neuroscience. These academic conflicts are not new to science, and we may remind ourselves of the teaching of Aristotle more than 2000 years ago to justify this conflict – when pupils at the *Lyceum* asked how he could differ from his master Plato, Aristotle had said *'Amicus Plato, sed magis amica veritas'* (Dear is Plato, but dearer still is truth). Ultimately in science, truth should prevail over the age-old beliefs.

Sources and Acknowledgments

Throughout this book I have cited references wherever appropriate, all from standard sources, not only to make my work authentic but to provide an inquisitive reader the opportunity to further expand his/her knowledge. The details of the references are supplied in the Reference list at the end of the book.

I should acknowledge that the following sources of information, in particular, were essential for me to develop the Molecular-grid Model. *Guyton's Textbook of Medical Physiology*, 11th edition (cited as Guyton & Hall 2008 in the text) is an esteemed source of the fundamentals of human physiology and is a must-read for all biologists interested in the human body. *Ganong's Review of Medical Physiology*, 24th edition (cited as Barrett et al. 2012) is a reputable textbook, which provides us with an in-depth knowledge of the issues in physiology. These books gave me invaluable insights into the subject of neuroscience.

In order to understand cell biology, it is essential to delve deep into the affairs of physiology of the neuron – after all, a neuron is a cell. *Becker's World of the Cell*, 8th edition (cited as Hardin et al. 2012) is an exhaustive textbook in the field of cell biology. It has certainly expanded my comprehension in the workings of neurons – especially regarding recent concepts about cell membranes. It is arguably the best book in the market in its field. *The Brain from Top to Bottom* (cited as Bruno 2002) is a respectable website from McGill University that exhaustively deals with all aspects of neuroscience and must be read by all biologists interested in this subject – it has many useful web-links too. *Britannica Guide to Brain* (Fine 2008) is an interesting book on the basics of the brain. Finally, *The Story of Philosophy* by Will Durant (Durant 2006) is a profound saga of the evolution of philosophical thought across generations – all the way from Aristotle to Professor John Dewey – and this book is meant for non-philosophers. From this source I have taken the metaphysical aspects of thought for the discussion at the end of the book.

Writing a book is different from publishing. Having acknowledged the indispensable repositories of knowledge, I must now identify my benefactors who helped this work see the light of the day. First of all, I must credit Natalie Farra, our Elsevier Neuroscience Editor, for her constant encouragement and advice, and for her enduring patience with my idiosyncrasies all though the production. I must thank Kristi Anderson, our most skillful Editorial Project Manager, for her indefatigable assistance in shaping the book. And my heartfelt thanks to the Elsevier publishers and staff who made this work possible. I feel it's my duty to thank our Production Manager at Elsevier, Chris Wortley, for his great diligence with this work and for gracefully accepting so many of my last-minute additions during production. I thank our copy editor Lynn Watt whose thorough work has added a professional glint to the book. I would also like to thank Professor M. Rameshwarudu, Dean and Professor of Physiology, SVS Medical College, Mahabubnagar for his constant academic support throughout this period. A good deal of criticism went on which has led to crucial changes – I should thank my colleague-friends Drs Srinivas Bandari, Venkatesh Ambarkar and Ramavara Prasad for their support. Finally, my more-than-customary thanks to my wife, Sathya, for her patience and good advice when it was most needed.

Krishnagopal Dharani

PART I

BASIC CONCEPTS IN NEUROBIOLOGY

CHAPTER

1

Functional Anatomy of the Brain

OVERVIEW

The human brain is the most sophisticated and complex product of biological evolution, and forms the structural basis of all human thoughts and actions. The human brain weighs about 1.5 kg – only three living species have brains larger than human beings, the elephant, the whale and the porpoise – however, in humans the weight ratio between the brain and the body far exceeds any of these.

The anatomy of the brain is better described in relation to its functions. Though there are significant advances in the understanding of human brain function, the knowledge is far from complete and the function of some significant portions of the human brain remains mysterious. The following description of the human brain may appear a little over-simplified, but it does serve as a useful guide for exploring the secrets of the brain.

GENERAL PLAN OF THE NERVOUS SYSTEM

Anatomical Considerations

The brain (Fig. 1.1) is the most important coordinating organ of the human body. It *receives stimuli* from the external environment and prepares the body to *respond suitably* to those stimuli. We will now study the types of nervous system and see how the complex human brain is structurally designed to receive the stimuli and to respond.

Types of Nervous System

Nervous system is conventionally divided into ***central nervous system*** (CNS) and ***peripheral nervous system*** (PNS). The brain and the spinal cord constitute the CNS, whereas nerves and other structures outside the

The Biology of Thought.
DOI: http://dx.doi.org/10.1016/B978-0-12-800900-0.00001-4

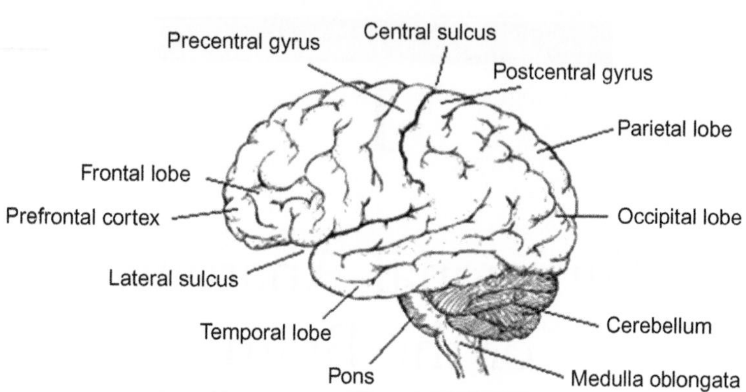

FIGURE 1.1 Lobes of brain.

CNS constitute the PNS. The PNS collects information from the outside world and sends it to the CNS; the CNS processes this information, and sends it back to the PNS for appropriate action.

Functionally, the nervous system may also be divided into the following three categories:

- Somatic nervous system
- Autonomic nervous system
- Enteric nervous system.

Somatic Nervous System

When a pin pricks the heel, for instance, the pain sensation is transmitted by the peripheral nerves to the spinal cord, upon which the spinal cord sends back signals through the nerves to the leg, instructing it to withdraw. This is an ***involuntary action***. As you can see, there must be two types of nerves – one to transmit signals *to* the CNS, and the other to pass messages *from* the CNS to the PNS (see Fig. 2.1). The former is called the ***sensory system,*** which carries the information to the CNS; and the latter is called the ***motor system,*** which transmits the CNS commands to the periphery. Take another example: when you look at a beautiful rose (this is a sensory input reaching the brain) you may try to pick the flower (this is a motor action initiated by the brain and transmitted to peripheral nerves). This is a ***voluntary action***. The system which controls *both* the voluntary and involuntary mechanisms is called the ***somatic nervous system,*** where there is a stimulating sensory input and a responding motor output. This looks simple enough.

Autonomic Nervous System

However, when you look at a horrifying scene, for example, your heart beats faster, you may sweat profusely, your hair stands on end and perhaps you may shudder – all this certainly without your conscious commands – the body's system has actually worked automatically. This is executed by another type of integrated nervous system called the ***autonomic nervous system*** (ANS), which regulates the body functions without your knowledge. Several life-saving mechanisms in the body are controlled by this system – when a sprinter runs fast his heart rate increases, as does his respiratory rate to keep up with his muscle's increased oxygen demands, and at the same time he sweats to regulate his body temperature – but for these adaptations, our sprinter would have been long since dead in his tracks.

Enteric Nervous System

There is another sort of nervous system called the ***enteric nervous system*** (ENS), which spreads along the gut and is *not* under the *direct* control of the CNS. When you eat, the food passes automatically down the esophagus and stomach to the bowels for digestion and absorption – this is coordinated by the ENS, and it is somewhat functionally integrated to the ANS. However, this system keeps functioning of its own accord, even without the supervision of the CNS and ANS; for example, even in deep coma (when CNS functions are interrupted) the gut still accepts food, digests it and propels it. Another example is that of the vagus nerves (nerves of the ANS to the gut) – even when they are cut surgically, as needed for duodenal ulcer therapy, the intestines function almost normally with the help of the ENS.

Types of Neurons

Sensory Neurons, Motor Neurons and Interneurons

In order to carry out the above functions, the nervous system employs three types of neurons (nerve cells; see Fig. 2.1). They are:

- Sensory neurons
- Motor neurons
- Interneurons.

In order to recognize the sensations reaching the brain (such as light, sound, touch, etc.), a set of special neurons are required, called ***sensory neurons***. So, to execute the desired actions (such as moving an arm or leg), another set of neurons called ***motor neurons*** are required. And to mediate between these two, a set of really versatile neurons called ***interneurons*** are needed. The interneurons need a special mention here. For instance, when a medical lab technician jabs a needle into your arm to collect blood, you do not withdraw your arm, right? At least not violently. In this case the sensations pass to the sensory neurons of the *brain*, and before they pass to the motor neurons they pass through the interneurons present in between the sensory and motor neurons, from where the message passes to different areas of the brain to carry out intelligent decisions before responding – they tell you that it is only a lab technician who has pricked you and there is no need to react. A needle prick in this case has caused the mind to make an elaborate analysis of the situation before allowing the body to act. However, the sensory neurons do send a little information to some motor neurons as well, and this may let you jerk your arm slightly.

Whereas if a needle pricks you unawares, it sends a shock wave across the body and you suddenly withdraw your arm. In this case the message passes to the sensory neurons of the *spinal cord*, which sends an instruction *immediately* to the adjacent motor neuron to withdraw the limb *reflexly* (***spinal reflex***). And there are interneurons in the spinal cord between sensory and motor neurons too, which send signals to many of the brain's other neurons, letting your brain know that a needle has pricked you and perhaps producing a myriad other somatic and autonomic effects such as wincing and producing a few tears. However, most of the interneurons keep the signals in check before reaching the motor neurons, otherwise each and every sensation would invariably result in an action akin to a puppet show – which is all the more unpleasant and disturbing (see Chs 2 & 5 for more detail).

Functions of Nervous System

The functions of the human CNS can be summarized into the following five categories:

- Sensory functions
- Motor functions
- Autonomic functions
- Involuntary functions
- Intelligent functions.

Sensory Functions

The sensory neurons can detect a wide variety of stimuli. The signals pass upwards to the brain along the nerves and reach neurons in the central nervous system (hence called *afferent* or *ascending pathways*; see Ch. 2). The sensory stimuli can be of two types – ***special senses*** are vision, hearing, smell, taste and vestibular function (sense of balance and spatial orientation, especially important if you want to walk on two legs), and ***general senses*** are touch, pain, temperature, pressure, vibration and proprioception (the sense of relative position of body parts in space – for example, this sense lets you know the position of your arm if you hold it up in the air, even with your eyes closed). The special senses arise from the specialized sense organs present in the head and neck and travel via ***cranial nerves*** to reach the brain (see 'Cranial nerves' below); for instance, the sense of vision arises from rods and cones in the retina of the eye and travels via the optic nerve, or the sense of smell arises from the sense organs of the nose and travels via the olfactory nerve, and so on. The general senses arise from various sense organs from the trunk and limbs, and transmit via ***spinal nerves*** to reach the spinal cord and thence to the brain (see 'Spinal nerves', below). The skin is the largest sensory organ in the body. It must be noted that all sensations from one half of the body are sensed on the *opposite* side of the brain – for example, when the *right* foot is pricked, the pain is perceived at the *left* hemisphere.

Before reaching the brain, the sensory system has to pass through various controlling centers located in the spinal cord, midbrain and in the subcortical regions (see below). These centers are collections of neurons called ***nuclei*** (singular *nucleus* – but not to be confused with a nucleus in the center of a cell). One such center is the *thalamus* (see 'Thalamus', below). These centers act as relay stations (or gateways) to modulate the information passing through them (sometimes the signals are passed and sometimes they are checked, as required). This is why not *all* the sensations we get from the external world reach our conscious attention all the time; for instance, there are so many minute sounds around us, such as the ticking of the wall clock, the slight buzzing of your computer fan, the constant 'touch' sensation from your back while sitting in a chair, all of which will not be perceived unless we pay particular attention, because they are controlled by these relay stations. These 'unimportant' sensations are abolished by these nuclei to facilitate better appreciation of more useful sensations.

Motor Functions

The motor neurons are the effectors that pass the signals along the nerves to the effector equipment, which are chiefly the muscles and glands (hence called *efferent* or *descending pathways;* see Ch. 2). In the above example of the pin-prick to the heel, the response to pain was to withdraw the leg – the pain impulses had passed up along the sensory nerve fibers and reached the appropriate neurons in the spinal cord, and from there the signals were routed to the motor neurons, which had sent signals along their nerve fibers to the skeletal muscles of the lower limb and resulted in a withdrawal reflex.

When you pick a rose flower with your fingers this is a *voluntary action*, and the motor commands descend from the *cortex*. These motor pathways are relayed to various nuclei before sending commands to reach the peripheral nerves. The nerve fibers from the motor cortex of the brain end on the second level of neurons present in the brainstem (*corticobulbar tracts*) or descend directly to end on the motor neurons of the spinal cord (*corticospinal tracts / pyramidal system*). All these tracts decussate (cross over) the midline to control the *opposite half* of the body.

There are a few other motor relay stations (e.g. *basal ganglia;* see 'Parkinson's disease' below), which receive motor fibers from the motor neurons of cortex and refine the movements of the muscles they supply (called the *extrapyramidal system*). The chief function of 'pyramidal' and 'extrapyramidal' systems is to enable us to execute fine and fractionated movements of the muscles involved in articulation of speech and movements of hands and body. This feature gives humans the prodigious capacity to use their muscles in a variety of ways – for instance, to use fingers with dexterity (e.g. to strum a guitar precisely); to use the speech muscles to generate sensible sounds (e.g. to sing a melodious song); and to use trunk muscles rhythmically (e.g. to dance to the tune).

Autonomic Functions

The autonomic nervous system (ANS) regulates the physiologic processes critical to survival and adaptation, and controls the internal environment (*milieu interior*) through its action on smooth muscle, cardiac muscle, secretary glands, etc. The effects of the ANS are ubiquitous and protective. The regulation of blood pressure, blood flow through organs, sweating, hunger, satiety, changes in body temperature, thirst and maintenance of circadian rhythms are some of the feats of the ANS. Though the actions of this system are automatic (hence the name), they are closely integrated to the somatic (voluntary) system.

The autonomic afferents ascend along the somatic sensory pathways to reach the CNS. The efferent (motor) pathways take an interrupted journey before they reach the effector organs. A single autonomic nerve may synapse with 15 to 20 neurons – and thus a single stimulus may give rise to a widespread response; for example, a single needle prick may result in profuse sweating all over the body, or a mere look at a spicy dish may make our mouths water.

Involuntary Functions

Some functions are involuntarily performed, such as breathing, digestion, heart beating, eye reflexes, etc., but some involuntary actions have voluntary control to a certain extent – examples are breathing, salivation, deglutition (swallowing), defecation, micturition (urination) and others. This becomes obvious if you consider the following examples: one can breathe rapidly at one's wish; one can defecate or urinate at one's convenience. These involuntary functions are basically looked after by the midbrain centers, but because they are bossed over by the cerebral cortex they have also some voluntary control.

Intelligent Functions

Finally, the brain is the seat for all the 'higher' intellectual functions of the human being. Memory, learning, innovation, intuition, creativity and the other mental faculties of man certainly have some anatomical correlates in the brain, but as yet they are not clearly defined.

Having outlined the brain's functions, the anatomical aspects of the brain can be described for better appreciation of its functional operations.

STRUCTURE OF THE BRAIN

The brain (*encephalon*) is a soft, pinkish jelly-like organ, which lies protected in the bony cranium. Anatomically and developmentally the brain is divided into three regions:

- The **forebrain** (prosencephalon)
- The **midbrain** (mesencephalon)
- The **hindbrain** (rhombencephalon).

The ***forebrain*** consists of the ***two cerebral hemispheres*** and the ***diencephalon***. Cerebral hemispheres constitute the major portion of the brain, whereas the diencephalon is situated inside the cerebral hemispheres. The ***midbrain*** lies midway between the diencephalon and the hindbrain. The ***hindbrain*** consists of the ***pons, cerebellum*** and ***medulla oblongata***. The midbrain, pons and medulla oblongata constitute a functional unit called the ***brainstem***, which forms an important portal for the to-and-fro traffic of the signals between the body and the brain. The brainstem carries out several important vital life-sustaining functions.

The normal human brain's weight is about 1400–1500 grams, but this is no measure of the intellect it harbors – Einstein's brain reputedly weighed about 200 g *less* than the average!

Three layers of *meninges* cover the brain and spinal cord. The brain is constantly bathed by a thin watery fluid called *cerebrospinal fluid* (CSF), which is secreted by the *choroid plexus* (a vascular tissue) present inside a system of interconnected cavities situated deep inside the brain called the *ventricular system*. The fluid thus formed from inside the brain comes out onto the surface of the brain through small openings near the medulla oblongata. This fluid circulates in the *subarachnoid space*, which then covers the entire surface of the brain and spinal cord. Normally, about 150 mL of CSF is present in dynamic circulation. Sometimes, developmentally, the openings at the medulla oblongata are blocked, which may result in accumulation of fluid inside the ventricles, a condition called *hydrocephalus* ('water-on-the-brain'). The brain receives its blood supply from the carotid and vertebral arteries. The brain is devoid of lymphatics. It must be noted that although the brain senses pain, the brain itself is insensitive to pain sensation.

CEREBRAL CORTEX

Each half of brain represents a cerebral hemisphere. Cerebral cortex consists of an outer layer of ***gray matter***, beneath which lies a thick mass of ***white matter***. The grey matter contains billions of neurons with their numerous dendritic ramifications and interconnections. The white matter is composed of axons arising from these neurons. The white matter appears white because the axons are ensheathed in wax-like myelin (see Ch. 2, p. 44).

The surface of the cortex is extensively convoluted (thrown into folds), which allows a much larger surface area of the brain to be accommodated in the limited space within the skull. The surface area of the human cerebral cortex is estimated to be approximately 2200 cm^2, which is about the size of a full newspaper page. The ridges are called ***gyri*** (singular – gyrus) and the furrows in between the ridges are called ***sulci*** (singular – sulcus). The pattern and size of gyri and sulci are somewhat inconsistent with individual variations, except for certain prominent deep furrows which are distributed with considerable regularity in all human beings. Three of them, called *central sulcus* (*or Rolandic fissure*), *lateral sulcus* (*or Sylvian fissure*) and *calcarine fissure*, are present on each cerebral hemisphere, which help in demarcating the lobes (Fig. 1.1).

Lobes of the Brain

The cerebral hemisphere is conventionally divided into six lobes:

- The frontal lobe
- The parietal lobe
- The temporal lobe
- The occipital lobe
- The insular lobe
- The limbic lobe.

The location of each lobe is better understood with the assistance of a picture of the brain (Fig. 1.1). These divisions roughly correspond to the functional aspects of the cerebrum. First the lobes are described along with their putative functions, and later the microanatomy is described in brief.

Frontal Lobe

This is the foremost portion of the brain, extending from the central sulcus to the front. The area just in front of the central sulcus is called the *precentral gyrus*, which forms the *primary motor cortex* (Brodmann's Area 4). This area contains a detailed topographical motor representation of the *opposite* half of the body, and controls all the skeletal muscle movements on that side. It means that Area 4 of left cerebral hemisphere controls the movements of the right half of body; and that of the right controls the left. A striking feature is that it represents *all* portions of the body, from head to toe, but the actual extent of representation of each part of the body varies greatly. For instance, the area of cortex representing the muscles of face, tongue and hand occupies a much larger area than to the area of cortex representing legs and toes – for the simple reason that the muscles of the face, tongue and hands need many more neurons for finer movement

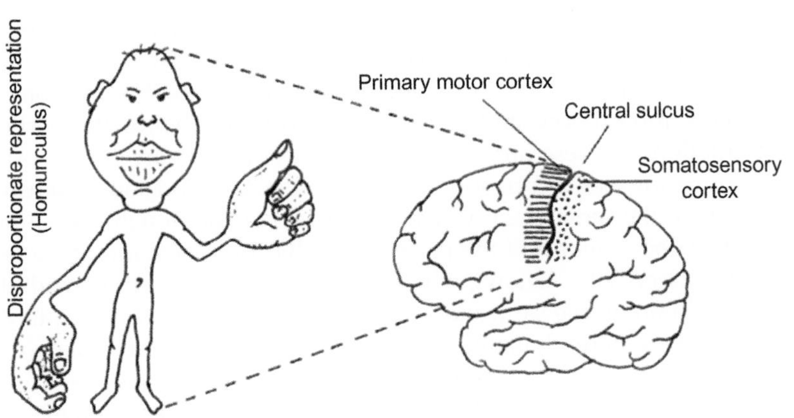

FIGURE 1.2 Homunculus – sensory/motor.

control, than for the control of leg and toes. When we illustrate this sort of representation on the precentral gyrus, it gives a grotesque shape of the human body called *motor homunculus* (Fig. 1.2). (A similar depiction can be made of the sensory cortex, called sensory homunculus, and both homunculi share strikingly similar features, see Parietal Lobe below.)

Other important areas of the frontal lobe are the *premotor cortex*, which is responsible for the control of precise direction of movement and its anticipatory behavior; *frontal eye field cortex* to control eye movements and gaze fixation; *supplementary motor cortex* to control complex tasks, temporal organization and retrieval of 'motor memory' (see Ch. 3, p. 62).

Broca's area (*motor speech area*; Areas 44, 45), a well-studied part that controls the muscles of speech (see Ch. 3, p. 73), is located here. Another important area, called the ***prefrontal cortex*** (PFC; Areas 9, 10, 46; Fig. 1.1), is responsible for the higher functions, such as intelligence, personality, foresight and planning, which are so characteristic of human beings. Needless to say that the frontal lobe has dense connections to the whole of the brain. Much research has been devoted to the PFC in the recent times to study its importance in short-term memory (see Ch. 3). Recent studies have shown that a specific region called ***dorsolateral prefrontal cortex*** is responsible for several higher intellectual functions like human reasoning, emotional response to distress and working memory. The prominence of prefrontal cortex in man explains his high forehead (unlike the receding foreheads of his cousins, the primates).

Parietal Lobe

The parietal lobe lies behind the frontal lobe. The *postcentral gyrus* (gyrus behind the central sulcus) constitutes the ***somatosensory cortex*** (Areas 1, 2, 3). This is where the whole *opposite* half of the body's sensory inputs (general sensations of touch, pain, temperature, etc.) is registered. The total body is represented in the *sensory homunculus*, similar to that of the motor homunculus; for example, the human thumb, lips and tongue have many more nerve endings than the toes and hence the cortical representations of these are much larger, whereas the part of the cortex representing the toes is correspondingly smaller. A second somatosensory area exists in the same vicinity where transient tactile (touch) sensations are appreciated. The rest of the parietal lobe has rich *association areas* (see 'Cortical connectivity', below), which connect the sensory cortex with the rest of the brain on the same side, apart from connecting this area to the opposite hemisphere, thalamus, brainstem and spinal cord. Apart from sensory functions, a significant portion of the posterior parietal cortex partakes in motor control of the body; thus, loss of primary motor cortex (Area 4) itself will not completely abolish motor activity.

Temporal Lobe

The temporal lobe is situated below the parietal lobe, beneath the lateral fissure and nearer the base of brain. Along with the frontal lobe, the temporal lobe is greatly expanded in humans, signifying the anatomical relationship of these areas to intelligence and memory.

The ***auditory cortex*** (Area 41) corresponds to the upper temporal gyrus, and the rest of the temporal lobe forms association areas. It is suggested that the temporal lobe is 'polysensory' in man because it integrates the auditory, sensory, visual and limbic functions. Certain advanced functions like face recognition is performed in the *fusiform gyrus* of dominant (usually the left) temporal lobe. A well-defined area called ***Wernicke's area*** (Area 22) is situated in this lobe on the dominant hemisphere, which is concerned with understanding written and spoken language (Ch. 3, p. 72; Fig. 3.10). It is interesting to note that in musicians the size of the auditory area is much larger owing to constant appreciation of musical tones (Barrett et al. 2012, p. 208).

Occipital Lobe

The occipital lobe occupies the back portion of the cerebral hemisphere, and houses the most important of the special senses – the visual function. The ***primary visual cortex*** (Area 17) is situated far back in the occipital lobe, surrounded by the visual association areas (Ch. 5, p. 103). The important feature of visuospatial perception in the execution of working memory is thought to be accomplished by these association areas. The subject of vision is discussed in some detail in Chapter 3.

Insula

The insular lobe lies deep in the lateral fissure, the function of which is largely unknown, but appears to be concerned with the appreciation of taste and perhaps also some aspects of language.

Limbic Lobe

The limbic lobe comprises a wide portion of cortex on the inner side of each hemisphere including parts of the frontal and temporal lobes. Conventionally, it includes cingulate gyrus, subcallosal cortex, parahippocampal gyrus and olfactory cortex. *Cingulate gyrus* (Figs 1.3 & 1.8) is situated in the medial aspect of each hemisphere. *Parahippocampal gyrus* is a part of the temporal cortex, and has the following chief components: hippocampus, dentate gyrus, subicular complex and entorhinal cortex. *Hippocampus* (Fig. 1.3) is the curled-up inner fold of temporal lobe, forming the most medial part of the parahippocampal gyrus, a cross-section of which looks like a sea-horse, hence the name (*hippokampos* in Greek = sea horse). Hippocampus and its associates are concerned chiefly with memory formation, and are described in Chapter 3.

The term ***limbic system*** is applied to a functional unit of the brain associated with instinctual behavior and emotions, and is composed chiefly of limbic lobe, amygdala, anterior thalamic nuclei, septum and possibly hypothalamus, habenula and parts of basal ganglia.

Amygdaloid Body (or Simply Amygdala)

Amygdala (Fig. 1.3) is an almond-shaped body situated in front of the hippocampus, and is considered a quasi-cortical structure. It is an important connecting part of the nervous system – connecting the cortical areas with the subcortical structures. It forms an essential part of the limbic circuit concerned with emotions. Amygdala is perhaps the most prominent functional unit of the brain involved in generation of fear. Apart from this, it is also concerned with instinctual reactions to the presence of food, rivals or sexual partners, or for protecting their offspring. All of these

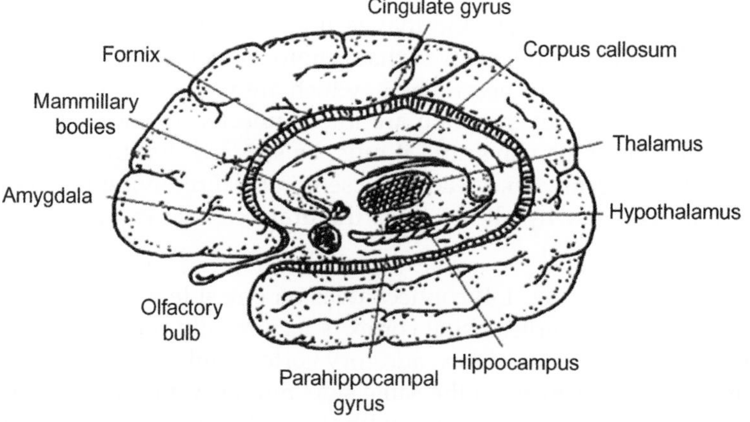

FIGURE 1.3 The limbic system.

emotions are important for survival. It has been found that the amygdala appears larger in neglected children who were emotionally deprived, obviously indicating that these children have an over-developed instinct for food and rivals.

Why Do Children Laugh or Cry More?

The amygdala in children is relatively well developed, whereas the cerebral cortical connections are less so. Amygdala is concerned with generation of emotions, whereas adult cortex can consciously suppress such emotions. Thus the immature cortex of a child cannot suppress amygdala – and this is the reason children cannot control expressions in the event of joy or sorrow. For example, adults can subdue their laughter when a man on the street trips over a banana peel, but children hardly can. Children also cry when they see strangers because their amygdala works without cortical rationalization.

Olfactory Pathways

It is worthwhile noting that sensation of smell has an ancient lineage. The olfactory bulbs contain the receptor neurons that receive the smell sensation from the nose and transmit to the olfactory lobe, which is a part of limbic system. The olfactory pathways are *directly* connected to the olfactory lobe, bypassing thalamus – perhaps the only sensory input that bypasses thalamic supervision before reaching the cortex is that of smell – while *all* other sensory modalities are directed first to the thalamus before reaching their respective cortical areas (see 'The thalamus' below).

Cortical Connectivity

It is important to understand that each lobe of the brain is not an isolated part but all lobes mutually exchange information constantly through an extensive network of nerve fibers (which are really axons of neurons, see Ch. 2). *All* lobes of the brain are interconnected to each other in *all* of the following three ways, and it must be stressed that these lobes are connected reciprocally – i.e., each lobe sending *and* receiving fibers at the same time. This is why all the areas the brain can act in unison to any given situation.

1. *Association fibers* (Fig. 1.4) connect areas on one side with areas on the *same* side; for example, visual cortex is reciprocally connected to the sensory cortex, motor cortex, auditory cortex, limbic lobe, prefrontal cortex and other areas on the same side. Similarly, the sensory cortex has dense connections not only with the motor cortex but with all other areas on the same side. Likewise all areas are interconnected.

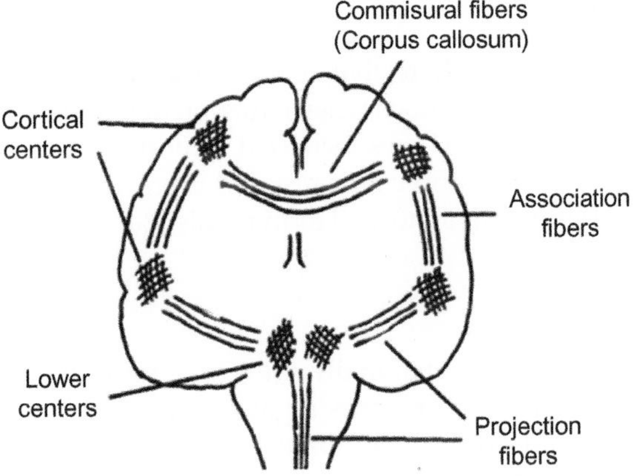

FIGURE 1.4 Cortical connectivity.

This ingenious arrangement has profound implications; for example, if you *hear* a loud sound, you *turn* your head so that you can *see* what has caused it – because the auditory cortex hears it, at the same time sending signals to the motor cortex to turn the head, the visual cortex to see and the limbic lobe to react – all in a jiffy. Another example: Broca's and Wernicke's areas are the main computing hubs of language processing, the former involved in speech production and the latter in understanding written and spoken language (a person needs both these areas to write a sentence, to understand its meaning and to say it out loud) – this coordinated linguistic accomplishment is only possible because both these areas are densely connected to each other (Ch. 3, p. 73).

2. ***Commissural fibers*** connect one area with that on the *opposite* side. The bulk of these commissural fibers run from one hemisphere to the other in a midline structure called the *corpus callosum* (Figs 1.4 & 1.8). It is because of this crossover mechanism that the right hemisphere controls the left half of body; the left hemisphere controls the right half. This allows an efficient cross-talk between the two halves of brain, letting one half know what the other half is doing.
3. ***Projection fibers*** connect the cortex with the subcortical areas such as thalamus, basal ganglia, midbrain, spinal cord, etc. Here also the connections are reciprocal – the cortex *sends* and *receives* fibers to the lower centers. The thick bundle of white matter (consisting of projection fibers) that is located in the subcortical region before it enters the diencephalon is called the *internal capsule* (Fig. 1.9). This area is of great clinical significance because *stroke* frequently involves this region, the result of which is paralysis of the opposite half of the body (*hemiplegia*).

Cortical Plasticity

It is worth mentioning here that most of the cortical areas described above can take up other functions when situation demands. For example, in blind persons visual cortex shows increased activity with touch and sound sensations, whereas deaf people may respond more quickly and accurately to moving visual stimuli than normal individuals (Barrett et al. 2012, p. 208).

Cerebral Asymmetry

Both anatomically and functionally, the two hemispheres are *not* mirror images. There are subtle differences regarding the size and shape of the sulcal pattern on both sides. These anatomical differences probably reflect their functional differences too. However, these differences vary from individual to individual. Left hemisphere is the dominant hemisphere in nearly 91% of the human population, making them right-handed. In general, it is observed that the greater the function of a particular region, the greater is its size and cellular complexity; for example, in right-handed individuals, the sensory and motor areas adjacent to the central sulcus appear larger on the left side, indicating that the right-handed preference needs a larger neural circuitry on the left hemisphere.

It is interesting to note that sometimes the same area on either side controls different actions. For instance, Brodmann Area 22 (encompassing Wernicke's area) on the left side is concerned with speech, whereas the right side is concerned with music; and *planum temporale*, a portion of the temporal gyrus, is larger on the left than on the right and is concerned with understanding spoken language (Barrett et al. 2012, p. 208).

However, it is generally accepted that the left hemisphere is predominant in verbal, linguistic and analytical functions. It is also considered to be important in mathematical skills and in verbal memory. The right hemisphere is more concerned with spatial perception, holistic 'gestalt' thinking, music appreciation and non-verbal memory (Fig. 1.5).

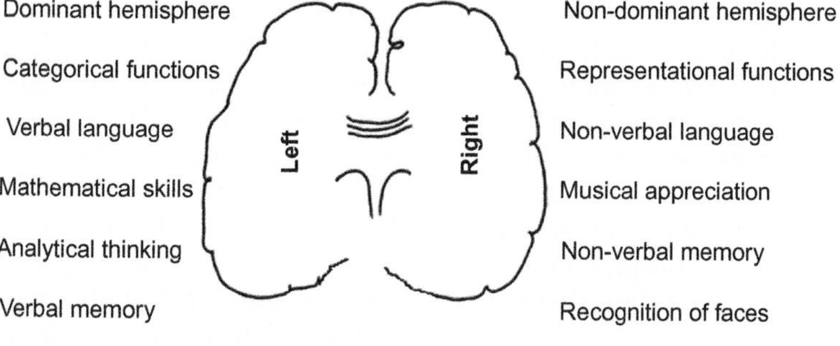

FIGURE 1.5 Cerebral asymmetry.

We can now say that the old adage, 'the left brain has nothing right and the right brain has nothing left' is not only false but is exceedingly uncharitable.

MICROSCOPIC ANATOMY OF CORTEX

Neuron

The human brain consists of more than 100 billion neurons, which intercommunicate extensively. The neuron is the basic functional unit of the nervous system, which performs all its physiological duties. There are many kinds of neurons which perform various specialized functions – such as sensory neurons, motor neurons, interneurons, etc. The neurons are electrically charged, and receive or transmit signals in the form of voltage potentials – these aspects of neurons are described fully in Chapter 2. The neurons form extensive neural circuits, estimated to comprise a length of hundreds of thousands of kilometers in the human brain.

Each neuron (Fig. 1.6) consists of a central body, called ***soma*** (pleural *somata*) and a number of membranous filamentary projections, called *neurites,* which emerge from the body. One of the neurites is much elongated and sometimes travels vast distances. This is called the ***axon*** (the so-called 'nerve fiber'). The rest of the neurites, called ***dendrites,*** are short and tentacle-like. Each neuron possesses about 5–7 dendrites but sometimes they are present in greater numbers. Characteristically, all dendrites branch extensively to form bush-like *arborization* around each neuron. Neurons communicate with each other by the way of junctions called ***synapses*** (Ch. 2, p. 45). These structures are described more fully in subsequent chapters as and when relevant.

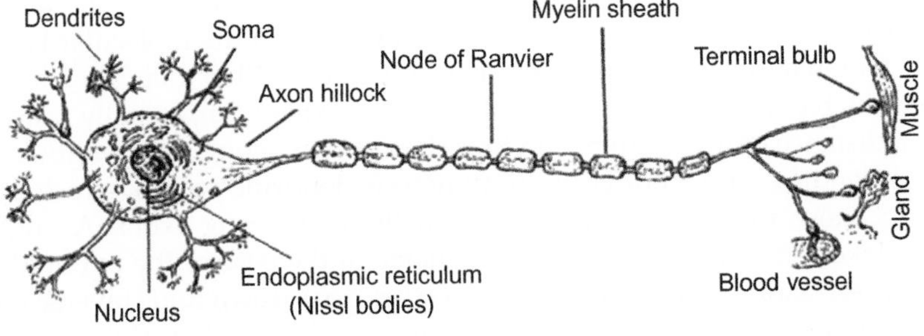

FIGURE 1.6 Parts of a neuron.

Neuroglia

The 'mass' that constitutes the brain comprises supporting tissue of the nervous system called neuroglia, and these cells outnumber neurons by 10–50 times.

They are responsible for creating a favorable environment for the neurons to function, although they have no neurological function themselves. Three types of neuroglia are described in the CNS – *astrocytes, oligodendrocytes* and *microglia*, which hold the neurons in neuropil and ensheath the axons by producing a wax-like protein-lipid material (called *myelin*) which wraps around axons, and electrically insulates them from surrounding axons, assisting in faster and better signal transmission (see Ch. 2, p. 44). The astrocytes are known to form 'gap junctions' among themselves through which they communicate freely with each other, and these are responsible for the regulation of synaptic microenvironment consisting of various neurotransmitters, ions and other metabolites.

Astrocytes and endothelial cells (blood vessel cells) form a selectively permeable layer around the blood vessel capillaries, called the *blood–brain barrier*, which is responsible for checking the passage of harmful substances from blood to brain, thus protecting the brain from the intrusion of various unwanted proteins and other harmful substances.

STRUCTURAL TYPES OF NEURONS

There are three main types of neurons in the cortex – the pyramidal cells, spiny granular (or stellate) cells and non-spiny granular cells.

Pyramidal Cells

By far the most prominent of the cortex neurons are the pyramidal cells, which are large and flask-shaped, about 10–80 μm in size. Their characteristic feature is the presence of a thick apical dendrite, which reaches the topmost layer of the cortex (subpial region) and ends up in a fine mesh of fibers that run parallel in the topmost layer, called the molecular layer (see Laminar organization of the cortex below). Other dendrites arise from the basal region of pyramidal cells and extend horizontally, reaching a distance of up to 1 mm, and branch profusely along their length. All these dendrites are studded with numerous dendritic spines, which are the locations for synaptic connections with neighboring axons. A single axon takes off from an axon hillock situated at the base of the pyramidal cell. These axons become myelinated as they pass down and traverse the white matter. Pyramidal cells are excitatory, and use either glutamate or aspartate as neurotransmitters (Ch. 2, p. 51). Generally, but not necessarily, these neurons are motor in function.

Spiny Granular Cells (or Stellate Cells)

These are the second most abundant neurons in the cortex. They are smaller (6–10 μm), but their dendrites show rich arborization (tree-like branching). Their axons ramify within the cortex, and they pass vertically down rather than horizontally. Their neurotransmitter is glutamate. These neurons are generally sensory.

Non-spiny Granular Cells

These form a group of heterogeneous (i.e., different types of) neurons, which constitute chiefly of interneurons (see 'Types of neurons' above). Their axons are confined to the gray matter. Several shapes are described, many with outlandish names such as basket cells, bipolar cells, chandelier cells, double-bouquet cells, Martinotti, neurogliaform and others. Many of them are inhibitory and use GABA (gamma-aminobutyric acid) as their neurotransmitter, but some use other neuropeptide chemicals like neuropeptide-Y, VIP (vasoactive intestinal polypeptide), cholecystokinin, somatostatin and substance-P (see Ch. 2).

Laminar Organization of Cortex

The cortex is characterized by the presence of a layered cellular distribution (Fig. 1.7). Typically, the human cortex has six layers (called *laminae*) and this type of cortex is called *neocortex* (Ch. 5, 'Locating perceptual

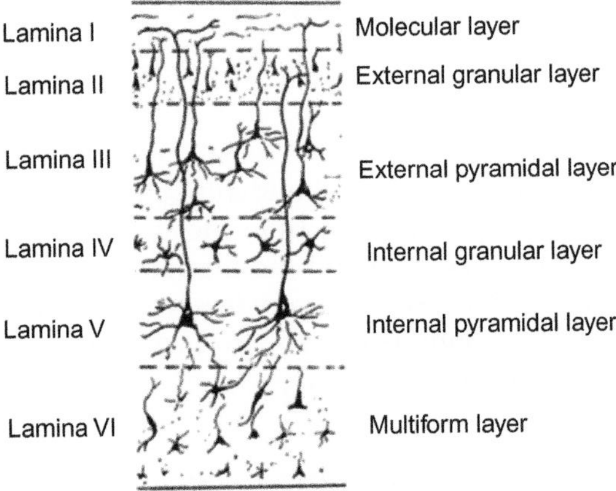

FIGURE 1.7 Layers of cortex.

neurons in the cortex'). Some areas of the cortex have fewer layers (called *paleocortex*, as in the cortex of olfactory lobe and the hippocampus).

- Lamina I. *Molecular (or plexiform) layer* – here the cells are sparse, only represented by a few scattered 'horizontal' cells. This layer contains fine leash of neurites derived from three sources: the dendritic arborization of almost all pyramidal cells of the cortex, the sensory axons of neurons of the cortex, and the axons of cortical interneurons.
- Lamina II. *External granular layer* – consists of small pyramidal cells and non-pyramidal cells.
- Lamina III. *External pyramidal layer* – small pyramidal cells.
- Lamina IV. *Internal granular layer* – spiny granular cells.
- Lamina V. *Internal pyramidal layer* – largest pyramidal cells.
- Lamina VI. *Multiform layer* – consists of a variety of cell types.

The distribution of this laminar pattern appears to have some functional significance, as different areas of the cortex show varying thicknesses of these layers; for instance, Layer V (pyramidal cell-predominant) is prominent in the motor cortex, whereas Layer IV (granular-cell predominant) is mainly associated with sensory areas. Further discussion on neocortex is undertaken in Chapter 5.

Brodmann's Areas

Although the human cerebral cortex is customarily described in lobes, there is an alternate classification wherein the cortex is divided in relation to its cytoarchitectural (i.e. microscopic) distribution, called Brodmann's areas. Brodmann was a histologist who painstakingly studied all the cortical areas and histologically divided them into different areas. This is a widely popular and useful topographical classification of the brain. They are numbered from 1 to 52, and they do bear some functional significance.

Rebirth of Neurons

At one time there was considerable debate about the rebirth of brain cells (neurogenesis). But now it is abundantly clear that in certain areas of the brain the neurons regenerate from the *stem cells* and the leading examples are the olfactory bulb and the hippocampus. The neurons in the olfactory bulb are replaced once every few weeks. Perhaps, the rebirth of neurons in the hippocampus is associated with the formation of new memories (Barrett et al. 2012, p. 287).

DIENCEPHALON

The diencephalon (Fig. 1.8) is a deep-seated part of the brain made up of several groups of nuclei situated amidst white matter. It consists of

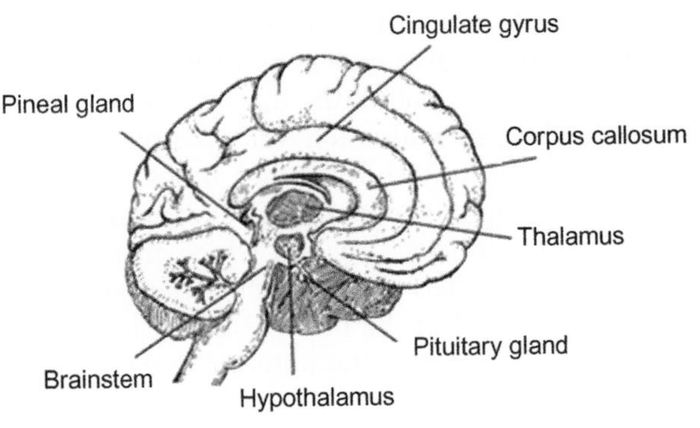

FIGURE 1.8 The diencephalon.

the *thalamus, hypothalamus* (including the *pituitary gland*), *subthalamus* and *epithalamus* (including the *pineal gland*).

Thalamus

The thalamus (Figs 1.8 & 1.9) occupies the central portion of the diencephalon. It is ovoid in shape, about 4cm long, and is formed by many groups of distinct ***nuclei*** divided into anterior, posterior, medial, lateral, pulvinar and reticular nuclei.

The thalamus is extensively connected to all areas of cerebral cortex above, and also to the brainstem and spinal cord below. The *thalamo-cortical neural pathways* send signals that arrive at the thalamus from various parts of human body to the cortex; whereas *cortico-thalamic pathways* send signals from the cortex to thalamus, thence to the rest of body. This means that it acts as a *'gateway'* between the brain and the body, and forms an important relay-station. Obviously the important function of the thalamus is to issue 'gate-passes' to these to-and-fro signals. Imagine the volley of signals that reach your brain each minute, but you also realize how many of them are actually *registered* by your conscious mind – the thalamus filters the unnecessary messages and allows only those signals that need attention. As an example, we are not aware of the constant ticking sound of the wall clock until we pay attention to it; nor are we conscious of the constant touch sensation caused by our clothes; or even, consider this – a sleeping mother wakes up at the slightest sound from her child, but she is not easily awakened by the infernal din of the street outside. Likewise, the thalamus does not allow our conscious mind to register so many other sensations that are not relevant in a given situation.

Perception of pain is another important thalamic function – it modulates the sense of perception, sometimes making pain 'pleasantly tolerable', at other times making a little pain just intolerably excruciating. This happens because the thalamus has rich connections to the limbic system, which controls our emotions – this enables the thalamus to have a flexible and contextual understanding of a situation; for example, heat is sometimes uncomfortable (scorching sun) and at other times may be pleasurable (e.g., a hot shower or a cup of steaming coffee). Another interesting phenomenon is the case of *'thalamic neglect'*, which manifests when some nuclei of thalamus are affected – this is characterized by a neglect of sensory inputs from one-half of the body resulting in a sort of dissociation. There is some evidence that the anterior thalamic nuclei regulate long-term memory (Ch. 3, p. 65).

Hypothalamus

Beneath the thalamus there is a small (0.3% of total brain weight) but important group of nuclei called the hypothalamus (Fig. 1.8), which is involved in the regulation of homeostasis (maintenance of a stable internal environment). It regulates fluid and electrolyte balance, food and energy balance, heat regulation, immune responses and also the reproductory functions. The hypothalamus controls body functions in two ways: 1) by directly influencing the autonomic system and achieving a neural response (through the nerves); and 2) by influencing the pituitary gland and achieving an endocrinal response (through the blood). For example, when you do strenuous exercise your pulse speeds up, you breathe rapidly (this is a neural response), and also your metabolism works faster, raising your temperature (this is an endocrinal response).

It secretes several 'factors' (i.e. chemicals) that influence the pituitary to release or inhibit the relevant hormones. *Somatostatin* (growth-hormone-release-inhibiting hormone), *thyrotrophin-releasing hormone, corticotropin-releasing hormone, neuropeptide-Y* and *neurotensin* are examples of such factors.

Circadian Rhythm

The suprachiasmatic nucleus of hypothalamus sets an endogenous circadian oscillator (an internal clock), which helps humans (and also higher animals) to maintain periodicity in their behavior. Our body clocks are set to function in accordance with external changes – such as day and night, periodical seasons and other such events (see pineal gland, below).

In general, it is said that stimulation of anterior hypothalamic centers produces parasympathetic effects, such as vasodilatation, sweating and uncontrolled body temperatures. On the other hand, stimulation of posterior regions of hypothalamus results in sympathetic overactivity like vasoconstriction, piloerection (goose-flesh and hair-raising) and shivering (which generates body heat through overactivity of muscles). Here are some more examples:

1. A fall of as little as 5% of blood volume may induce the secretion of vasopressin from the posterior pituitary, causing an urge to drink water. The volume receptors are situated in the great veins of the neck, which sense this depletion of pressure and send signals to the hypothalamus via vagus and glossopharyngeal nerves through the *tractus solitarius*, inducing a thirsty feeling. Restoration of this volume promptly quenches our thirst.
2. A *'feeding center'* has been identified in the hypothalamus, injury to which depresses appetite, whereas stimulation results in voracious appetite. However, another region called *'satiety center'* is also identified, which depresses the appetite. All these functions are mediated by glucose receptors situated in these centers.
3. When an individual hears 'shocking news', they are likely to collapse because of the sudden generalized dilatation of blood vessels resulting from autonomic disturbance induced by the hypothalamus.
4. As soon as someone perceives the smell of their favorite dish from the kitchen, not only does their mouth water, but gastric acid is also released into their stomach, inducing a sense of appetite. This is because when the stimulating smell reaches the olfactory lobe, it sends urgent signals to the limbic lobe, which in turn sends information to the hypothalamus. The hypothalamus works with the autonomic system to increase salivation and acid output in the stomach (see Papez circuit, Ch. 3).

Pituitary Gland

The pituitary (Fig. 1.8) actually consists of two separate glands clubbed together – *the anterior pituitary* and *the posterior pituitary* (intermediate lobe, seen in lower animals, is rudimentary in humans). The anterior lobe secretes seven hormones – *thyroid-stimulating hormone, adrenocorticotropic hormone, luteinizing hormone, follicle-stimulating hormone, prolactin, growth hormone* and *β-lipotropin*. The posterior lobe secretes *oxytocin* and *vasopressin*. These anterior pituitary hormones act on other endocrine glands like thyroid, ovary, adrenals and testis to stimulate them, and are responsible for the overall growth, metabolism and reproduction of the human body. Growth hormone deficiency results in *dwarfism*, where the child has a stunted growth, whereas its overproduction leads to *gigantism*.

HPA Axis

Here it is worth mentioning that the body has the ability to adapt itself to the changing environment owing to rich connections between glands – we sweat a lot in hot weather to lose heat, and shiver in cold to generate heat; our body metabolism has to work faster during exercise and is slower at rest; the heart has to pump more blood during exercise, and so on. For such mechanisms to operate the hypothalamus has to act upon the pituitary, which in turn acts on the adrenals to increase or decrease the catecholamine release into the blood. This is called the *hypothalamic–pituitary–adrenal axis* (HPA axis).

Pineal Gland

The pineal gland (Fig. 1.8) is an 8mm long, reddish-grey organ, situated in the posterior region of the thalamus. It is an endocrine gland that is involved in the regulation of circadian rhythm of the body. *Melatonin*, the hormone of the pineal, shows variations in its blood concentration levels through the day – darkness increases melatonin levels, which helps in maintaining metabolism at a low ebb during nights; whereas during the day its levels in blood drop, thus stirring up metabolism to be ready for the fray. Melatonin is also thought to be involved in processing long-term memory. Even before puberty, the pineal gland starts its involution and calcium deposits (*'brain sand'*) start forming in its body. Medication with melatonin is sometimes used to alleviate the symptoms of 'jet lag' after intercontinental flights.

BASAL GANGLIA

The basal ganglia (Fig. 1.9) consist of a group of large subcortical collection of neurons, located between the cerebral cortex and the thalamus. The chief nuclei of basal ganglia are caudate nucleus, putamen and globus

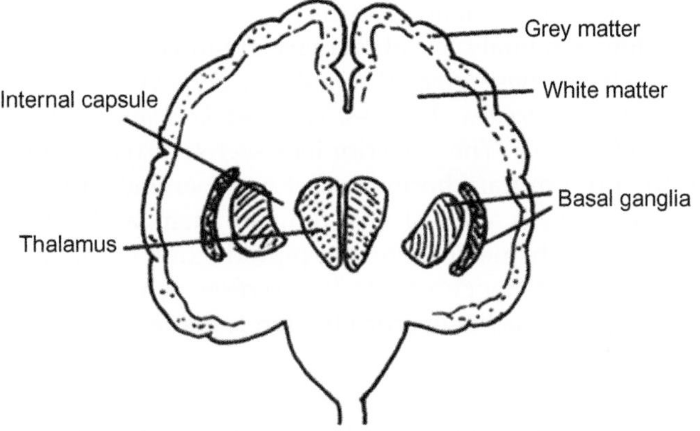

FIGURE 1.9 Basal ganglia.

pallidus. Above they are connected to the cerebral cortex, and below to the thalamus and brainstem nuclei. Amygdala, once considered a part of basal ganglia, is now considered a functional part of the limbic system.

The basal ganglia have a role in the control of movements and the motor behavior appropriate to a given situation, and facilitate necessary movements and inhibit unnecessary movements – thus, they are responsible for the execution of appropriate movements and suppression of inappropriate movements. In short, they 'smooth' the voluntary motor movements. The motor messages from the cerebral cortex reach the caudate nucleus and putamen first (these two collectively are called *neostriatum*), and thence to globus pallidus and *substantia nigra* (a part of the midbrain). The basal ganglia refine these signals and send them down to the spinal cord. This allows humans to execute skilled movements such as threading a fine needle. Another important function of basal ganglia is the facilitation of *associated movements*, such as swinging the arms while walking, facial expressions during emotions and speech, etc. Basal ganglia are concerned with initiation of voluntary action by converting an abstract thought generated in the cortex (Barrett et al. 2012, p. 245). A defect in these mechanisms is classically seen in Parkinson's disease.

Parkinson's Disease

Parkinson's disease is the most common affliction of basal ganglia that occurs due to the degeneration of dopamine-producing neurons of substantia nigra. When cells in the substantia nigra degenerate, the supply of dopamine (Ch. 2, see p. 50) is depleted. Dopamine has a dual effect on the basal ganglia, some excitatory and some inhibitory. In parkinsonism the overall effect is the inhibition of the motor thalamus, which results in slowness of movement, rigidity of muscles, tremors (shaking of limbs) and gait problems. Parkinsonism causes *akinesia* (slow voluntary movements) or *dyskinesia* (involuntary purposeless movements). Face becomes expressionless and the gait becomes slow.

The medical treatment of this disease lies in replenishment of dopamine (e.g. by administering *levodopa*), but in extreme cases neurosurgery may become necessary. In the past, thalamus or globus pallidus were the choice targets for surgery, but with the advent of stereotactic techniques ablation of subthalamic nucleus has become the preferred treatment. An alternative approach is the surgical insertion of 'pacemakers' into the basal ganglia (*deep brain stimulation*).

BRAINSTEM

The part of the brain which connects the cerebral hemispheres with the spinal cord is the brainstem (Fig. 1.1). It consists of *midbrain*, *pons* and *medulla oblongata*, and houses several life-supporting centers for the maintenance of vital functions such as breathing and blood pressure. The brainstem also has an ill-defined collection of neurons spread along its center, called the *reticular activating system* (RAS), which is of paramount importance in the development of consciousness, arousal, sleep regulation, and maintenance of muscle tone and posture. Reflex actions, such as swallowing, coughing, sneezing and vomiting, are located in the medulla oblongata. The nuclei of almost all cranial nerves lie in the brainstem. The cerebellum is connected to the brainstem through its peduncles.

Brainstem lesions cut off sensory input into the brain, at the same time blocking the brain's control over the body – and the results are often catastrophic, leading to a state of cortical disconnection called *decerebration*. In patients with deep coma who are on artificial respiration for a prolonged period, the attending physician has to make an accurate diagnosis of ***brainstem death***, commonly known as *'brain death'*. Ascertaining brain death is of paramount importance in modern clinical practice for the following two reasons:

1. To allow the physician to make a decision regarding withdrawal of artificial ventilatory support for life and to declare the patient dead (with legal implications)
2. Perhaps more importantly, brain death permits the removal of organs such as cornea, kidneys, liver and heart for donation.

CEREBELLUM

The cerebellum (Fig. 1.1) weighs only about 20% of the cerebral cortex, but its surface is so extensively furrowed that its total surface area comprises approximately 75% of that of the cerebral cortex. And its cortex is so densely packed that it actually contains more neurons than the cerebral cortex! The cerebellum forms something like an Elizabethan ruff around the brainstem, sitting astride the sensory and motor centers; in fact, the cerebellum coordinates all the motor centers of the brainstem.

Parts of Cerebellum

Functionally, the cerebellum is divided into three parts:

1. The vestibulocerebellum, concerned with equilibrium and learning-induced changes (anatomically represented by *flocculus* and *nodulus*).
2. The spinocerebellum (anatomically represented by *vermis* and *medial lobes*), concerned with proprioception (awareness of position of various parts of the body in space and their movement – if you hold your arm up in the air, the cerebellum lets you know its position

even with your eyes closed). This portion of cerebellum also has a 'copy' of the general motor plan derived from the cerebral cortex, which assists in coordinating the body movements.

3. The *neocerebellum* (or *'cerebro-cerebellum'*, anatomically the lateral cerebellar hemispheres) is concerned with complex movements, and this portion represents, phylogenetically, the most advanced part and is maximally developed in man. What you see prominently of cerebellum in human brain is this portion. It cooperates with the cerebrum to plan and organize voluntary movement.

The white matter of the cerebellum consists of four major *deep nuclei* by which the cerebellum connects with the brainstem and the rest of the CNS.

Cerebellar Circuits

The main input to the cerebellum comes from the *climbing fibers* of the *olivary nucleus*, through which the position sense from all over the body reaches the cerebellum. Though different skilled actions are executed by the prefrontal *cerebral* cortex, when such actions are repeated a number of times the learning task is shifted to the cerebellum, and these actions can then be executed without conscious control (for example, riding a bicycle or driving a car). This is 'motor memory' (Ch. 3, p. 62). Lesions of the olivary nucleus destroy this ability. The second input is from the *mossy fibers* which relay inputs directly from the cerebral cortex. These mossy fibers end up in complex synaptic groups called *glomeruli* (not to be confused with glomeruli of the kidney).

The climbing fibers and the mossy fibers synapse with the cerebellar Purkinje neurons extensively. These fibers exert an excitatory effect on the Purkinje cells, whereas the fibers from the granular cells inhibit the Purkinje cells. All the fibers from the Purkinje cells are consistently inhibitory to the deep cerebellar nuclei, whereas the output of the deep nuclei themselves is always excitatory to the brainstem nuclei. Thus the effect of inhibition of Purkinje cells is actually excitatory to the brainstem. In essence, this circuit is concerned with precise control of the cerebellar cortex over the deep nuclei, brainstem and thalamus. But the complete electrical activity that goes on in the cerebellum is not fully understood.

Examples of Cerebellar Activity

The cerebellum is the major motor center of the brain and is responsible for coordinated motor action. For example, when you bend your arm the biceps muscle has to *contract*; at the same time, unknowingly, the triceps (muscle of the back of arm) also has to *relax* (to stretch) to enable the limb to bend. These coordinated muscle actions and the general maintenance of posture is looked after by cerebellum. Man's unique capacity for bipedalism (walking on two legs) needs efficient cerebellar coordination.

The cerebellum is concerned with anticipatory action to maintain the upright position of human body. You must have noticed a tight-rope walker carrying a weight – they must balance in such a way that the gravity of the weight falls within the leg support over the rope, which is a real trick – and these complicated maneuvers are orchestrated by the cerebellum. Interestingly, this sense of position is informed to the cerebellum by the labyrinthine apparatus of the ear.

Signs of Disease

Humans with deranged cerebellum will not show any abnormality as long as the body is at rest, but when they move, the muscles show pronounced incoordination between the trunk and limbs. This results in a characteristic abnormal gait called *ataxia* (a sort of drunken gait), and a speech defect known as 'scanning speech'.

SPINAL CORD

The spinal cord lies in the vertebral column, and is continuous with the medulla oblongata. In the human being, 31 pairs of sequentially arranged spinal nerves carry the sensory inputs from the trunk and limbs to the spinal cord and thence to the brain; the spinal nerves also contain motor fibers to carry the commands from the spinal cord to the body. When a particular sensory input (e.g., a pinprick to finger) demands an urgent response, the spinal cord commands the motor nerves to act immediately. This is a *spinal reflex*. Of course, thoughtful and purposeful movements are performed by the brain.

The gray matter of the spinal cord is situated in its center (in contrast to the brain where grey matter forms the outer cortex). The nuclei of the afferent (sensory) fibers are located on the back portion of the spinal cord, whereas the efferent (motor) fiber's nuclei are located on its front. Thus the sensory roots (also called *dorsal roots*) of the spinal cord 'enter in' from its back, and the motor roots (also called *ventral roots*) 'emerge out' from its front. Dorsal roots and ventral roots join together to form *spinal nerves*, from where *peripheral nerves* emerge.

PERIPHERAL NERVES

Peripheral nerves (colloquially, simply called *nerves*) are bundles of axons running to-and-fro between the CNS and the body. Two types of peripheral nerves emerge from the CNS, as described below.

Cranial Nerves

Cranial nerves arise directly from the brain and supply the special sensory organs and general senses from the head and neck. The cranial nerves may be sensory or motor or of mixed type. There are 12 pairs, each designated with a Roman numeral, and some with exotic names:

I – *olfactory nerve,* which is concerned with the perception of smell
II – *optic nerve,* obviously the visual nerve
III – *oculomotor nerve,* which moves the eyes
IV – *trochlear nerve,* which also moves the eyes
V – *trigeminal nerve,* which is concerned with sensations of face
VI – *abducens nerve,* which also moves the eyes
VII – *facial nerve,* responsible for facial expressions
VIII – *vestibulocochlear nerve,* concerned with hearing and balance
IX – *glossopharyngeal nerve,* receives taste sensations from the tongue
X – *vagus nerve,* supplies the heart, lungs and intestines
XI – *accessory nerve,* supplies some muscles of the back and neck
XII – *hypoglossal nerve,* concerned with movements of the tongue.

Spinal Nerves

Spinal nerves arise from the spinal cord, and supply the rest of the body. Through the spinal nerves the CNS receives information and controls the actions of the trunk and limbs. They are formed by the union of dorsal and ventral roots. Thus, all the spinal nerves are composed of both sensory and motor fibers (mixed nerves). There are eight cervical, twelve thoracic, five lumbar, five sacral nerves and one coccygeal nerve (all in pairs), each one emerging out of their corresponding intervertebral foramina (bony openings) of the vertebral canal.

With this brief account of the anatomy of brain, we proceed to Chapter 2, to understand the unique physiological aspects of the neuron.

CHAPTER

2

Physiology of the Neuron

OVERVIEW

A neuron is the fundamental unit of the nervous system, the essential function of which lies in its superb electrical activity. This electrical activity enables it to perform its prestigious task of signal generation and transmission. Different types of neurons perform different activities, but all of them have certain general characteristics. A brief discussion of these general aspects of the neuron is presented here. A more detailed description of some of the physiological aspects of the neuron is given in subsequent chapters, where that information becomes more appropriate.

GENERAL PLAN OF NERVOUS SYSTEM

Physiological Considerations

The human brain receives a multitude of stimuli, not only from the external environment but also from its internal (bodily) parts.

External Stimuli

The human body is constantly bombarded with various ***stimuli*** from the external world. These stimuli are received by peripheral sensors called ***sensory receptors*** (or simply receptors), which generate ***sensations**** and send them to the brain for analysis and further action. The senses are of two types – special and general. The *special senses* are vision, hearing, smell, taste and vestibular function (sense of balance); the *general senses* are touch, pain, temperature, pressure, vibration and proprioception (sense of position; see Ch. 1, p. 6). The peripheral receptors are connected to the

* Throughout the book the sensory signals that arise in the peripheral receptors, such as eyes, ears, skin, etc., are called *sensations* for the lack of a better term. This is to differentiate sensations from *perceptions*, which are the signals that are perceived in the brain.

The Biology of Thought.
DOI: http://dx.doi.org/10.1016/B978-0-12-800900-0.00002-6

nerves, either cranial or spinal (see Ch. 1, p. 29), which take these senses to the brain. The brain responds by sending back messages to the *effectors* (i.e. body parts such as muscles and glands) through the motor system.

Internal Stimuli

Internal sensations are generally *visceral sensations* produced by various body parts and internal organs, ranging from specific sensations like hunger and thirst to vague sensations like dull pain, distension, lassitude etc. Life-threatening events, such as changes in blood pressure, pH or electrolytes, also send signals to the brain. The function of the brain is to coordinate and regulate the internal milieu of the body. The internal bodily stimuli reach the central nervous system (CNS) in two ways: 1) most often they take the form of hormones and other chemicals circulating in the blood, thus reaching the brain through the bloodstream, acting chiefly on the hypothalamus; 2) some internal sensations reach the brain directly by way of neural circuits, chiefly through cranial nerves.

The first type of response is a relatively slow process; for example, starvation stimulates adrenaline production, which releases glucagon from pancreas into the blood, this in turn induces gluconeogenesis in the liver (a process which makes up the levels of blood glucose in starvation) – all these sequential events do take some time to show their effect. However, when an urgent action is needed, the body relies on neural circuits, such as when the blood oxygen levels drop, the *carotid body* is stimulated due to alterations in the blood pH; this information is quickly relayed to the brainstem through glossopharyngeal nerve, and this immediately hastens breathing to restore the O_2 levels – this works on an emergency basis. We do not go into the details of these internal sensations in this book, but rather we will study the peripheral stimuli only.

Signals That 'Pass Up'

The body receives each type of stimulus, such as pain, touch, light, sound, etc. through specialized nerve endings (called *receptors*) present in appropriate locations in the body such as skin, eyes, ears, and so on (Ch. 5, p. 94). The chief function of these receptors is to convert the stimulus into an electrical signal; for example, pain is a mechanical stimulus and this has to be converted into an electrical signal in order to transmit along the nerves (because nerves can transmit only electrical signals but not any other); if the stimulus is light, the rods and cones convert light rays into electrical signals; if the stimulus is a sound wave the 'hair cells' in the ear canals convert them into electrical signals, and so forth. Thus, all the stimuli – whether they are from eyes, ears, nose, skin or any other sensory organ – *have* to be first converted into electrical signals before they are transmitted to the CNS. This process of conversion of one type of energy into another is called *transduction*. Hence, these receptors are called ***biological transducers***. It must be understood that the only way by which a stimulus reaches the CNS is by the way of an electrical signal. Thus, the *central tenet* in neuroscience is:

> Neurons understand only the language of electricity

How does this transduction take place? We will take vision as an example. When light rays impinge upon the photoreceptors (rods and cones) of retina, they get excited because the light energy incites ionic currents across their membranes. Because ions are electrically charged, these ionic disturbances create tiny voltage fluctuations, called ***membrane potentials*** (MP) across receptor's membranes; many of these MPs join together and gain strength to generate large voltage potentials, called ***action potentials*** (AP). The APs can now pass along the axons, and are commonly called ***nerve impulses***. The axons which carry nerve impulses are called the ***nerve fibers***. Thus, the peripheral nerves contain nerve fibers which carry APs. These nerve impulses are now ready to be carried to the neurons in the cortex.

Similarly, for an odor to be sensed, the smell-carrying chemicals must first dissolve in the thin mucus covering the smell receptors in the nose, which then excite their receptor's membrane to 'fire' APs, and these nerve impulses are carried along the axons that carry these sensations to the olfactory cortex where the sense of smell is appreciated. In the case of touch and pressure sensation, the receptors are the specialized nerve endings situated in the skin which contain sodium channels that get excited on slight touch or pressure, converting these mechanical stimuli into electrical potentials, and thus can be transmitted along their axons to reach the somatosensory cortex. Likewise, all sensations have to be transduced into electrical signals before reaching the CNS.

The impulses from the receptors usually pass several stages of neurons before reaching the cortex. One neuron is connected to another neuron until the signal reaches the final destination (Figs 2.1 & 2.8). The axons carrying sensory impulses generally end on ('end on' =connect to) the other neurons – this junction is called a ***synapse*** (see the summary of signaling activity of neurons, below). At the synapse the voltage potentials from axons are transferred to the succeeding neurons. The axons from these neurons end on the neurons that are connected further up, and this chain continues until the signals reach up to the neurons of the cortex.

The neurons described above are *excitatory*. Some other times, the neurons are *inhibitory* – they check the passage of signals further up. The difference

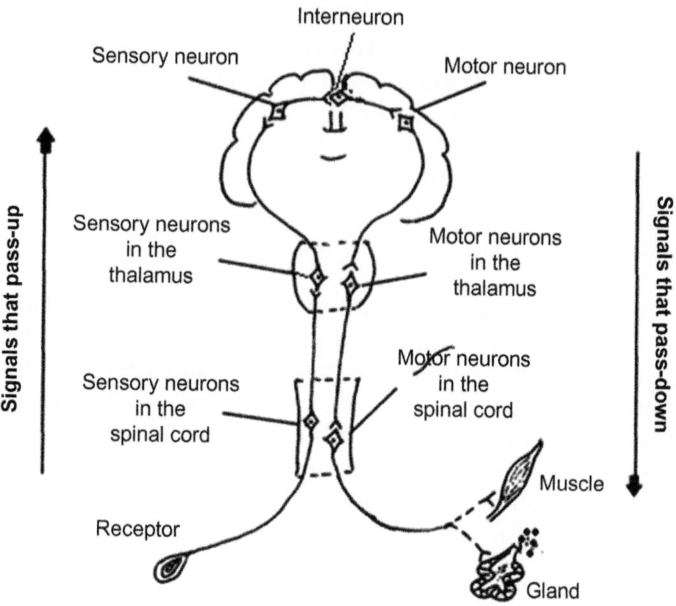

FIGURE 2.1 Signals that 'pass-up' and 'pass-down'.

between excitatory and inhibitory neurons lies chiefly in the neurotransmitters (NTs) they contain (see examples of NTs, p. 49–52, below).

Interneurons

The sensory neuron generally projects its axon to the *interneurons* (see p. 6), which are mediator neurons connecting sensory to motor neurons. You may wonder why sensory neurons do not connect directly to the motor neurons. The answer is obvious – if they passed messages directly to their motor counterparts, each and every time a stimulus is passed from the sensory neuron the motor neuron would *have* to respond and what would be the result? It would become something of a doll in a puppet show – every single pull at the string resulting in an awkward action. There must be a way by which you can have a control on your motor actions for a given sensory input. Thus, these versatile interneurons act like a ***switching mechanism*** in the circuit. Interneurons are further discussed in Chapter 5 (see p. 101).

A majority of the interneurons appear to be inhibitory so that they can abolish incoming sensations. However, some interneurons appear to augment sensations by spreading the sensory stimulus to many other motor neurons to show a more pronounced effect – this property is especially remarkable in the autonomic system.

Signals That 'Pass Down'

The signals are finally allowed to reach the motor neurons, which are then excited. The nerve impulses from motor neurons travel along their axons to reach the succeeding neurons until they reach their destination – be it another motor neuron, a muscle or a gland (Fig. 2.1). This gives varied responses; for example, skeletal muscles contract or relax (so that a person can move their limbs, turn their head, etc.), smooth muscles contract or relax (which causes changes in almost all organs in the body such as blood vessels, lungs, intestines, eyes, etc.), or glands are stimulated to release secretions or inhibited to dry up (so that the mouth may water or a body may sweat, etc.).

Complexity of Connections

In fact, the above description is a far too simplified picture of the real structure of the nervous system. The signal transmission in the nervous system is immensely more complex because of the extensive interconnections between the neurons. Each neuron has synaptic contact with hundreds, sometimes thousands, of the neighboring neurons because of extensive axonal collateralization (branching) and dendritic arborization. As the signals ascend, a single stimulus can excite many neurons in series to produce a big response. On an average, each neuron engages itself in about 2000 synapses and as there are 10^{11} neurons in the CNS, it can be calculated that there are about 2×10^{14} synapses (Barrett et al. 2012, p. 120). Thus the effect is that a single stimulus can produce a really huge response. Not only that the response is large, it is also immensely diverse. We will not go into the details of these connections as this is outside the scope of this work.

Now we will see how the neurons are electrically excited.

MECHANISM OF NEURONAL ELECTRICAL ACTIVITY

In the above discussion, we have been saying that the neurons and the nerves conduct *electrical* impulses; but, in fact, all these electrical signals are manifestations of *electrochemical events*, as we shall see now. And all these electrical signals can be understood by studying the ionic events that occur around the *neuronal membrane*.

The Electrical Signal

In a nutshell, a living neuronal membrane (as a matter of fact, the cell membrane of *every* cell in the body including neurons) is surrounded on

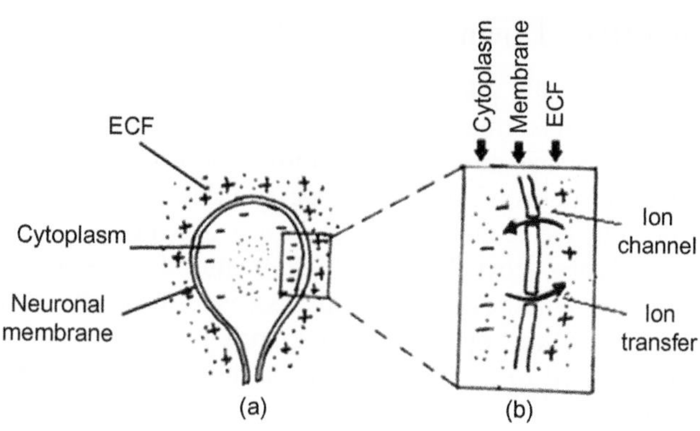

FIGURE 2.2 Neuron's ionic distribution.

either side by various electrically charged ions, but there is a potential difference in the concentration of ions on either side of the membrane. This difference in ionic concentration is responsible for the generation of electrical signals. Now we will see the finer details.

The cell membrane is surrounded by extracellular fluid (ECF) on its outer side and cytoplasm on its inner side (Fig. 2.2a), thus forming a three-layered functional unit (ECF–membrane–cytoplasm) that is crucial to understanding the electrochemical activity of a neuron. The ionic concentration of ECF greatly differs from that of cytoplasm, and this results in a constant transfer of ions across the membrane (Fig. 2.2b). This transfer of electrically charged ions generates tiny voltage potentials, called membrane potentials (MPs). It must be understood that not only neurons but all cells (including *all non-excitable* cells of the body like liver cells, fat cells, etc.) generate MPs, and these potentials are continuously produced across their membranes at all times, while the cells are alive.

However, the magnitude of these MPs is much greater in the case of neurons, which confers them peculiar electrical properties. In neurons these MPs are called ***resting MPs***, which usually bear a value of about −60 to −80 millivolts (mV). When neurons are excited the electrical properties of their membranes change, resulting in the generation of different types of nerve signals. A general mechanism of their production is discussed below.

Polarization, Depolarization and Hyperpolarization

Normally the interior of the neuron (cytoplasm) has an excess of negative charge, whereas the exterior (extracellular fluid) has a positive charge – i.e. inside of the membrane it is *negative* and outside it is *positive* – this is the normal ***polarized state*** (Fig. 2.2b). The positive charge outside the neuron is chiefly contributed by the sodium ions (Na^+) in the ECF, and the negative charge inside the neuron is mainly contributed by cytoplasmic proteins and other macromolecules inside the cell (Hardin et al. 2012, p. 368). Cytoplasm is also rich in potassium ions (K^+), which play a significant role in electrical activity. However, even at rest the cell membrane is 'leaky' to some extent, so that the ions pass both ways, and as time passes there is a tendency for this system to achieve a state of equilibrium when the transfer of ions comes to a standstill. But this state of equilibrium is not compatible with a cell's life. The neuron, in order to be in an excitable state, *must be* in a polarized state, because a state of equilibrium does not generate any potential. A cell at equilibrium is a dead cell (Hardin et al. 2012, p. 124)! Thus a constant 'maintenance job' in the form of ionic back transfer is required to keep up a steady polarized state. This ionic back transfer is executed by the ***sodium-potassium pump*** (i.e. **Na^+/K^+ -ATPase**), which constantly *pumps in* leaked-out K^+ ions *into* the neuron and *pumps out* leaked-in Na^+ ions (Fig. 2.2b).

Cell membranes are highly specialized structures and as they allow only specific ions across them, this property is called ***selective permeability*** (see Ch. 7, pp. 126). To enable this feat, membranes possess pores called *ion channels*. Ion channels are *transmembrane proteins* with 'pores' inside them, allowing ions to pass through. There are several types of ion channels, one each to conduct only a particular ion – *sodium channels* conduct only Na^+ ions; *potassium channels* conduct only K^+ ions; *chloride channels* conduct only Cl^- ions, and so on.

The ***neuronal membrane*** also houses these channels, but in a far *greater* number and perhaps arranged in a more organized fashion, and this enables them to transmit signals with profound intensity and at tremendous speeds. Swift ionic fluxes across the membrane can be modified and regulated by altering the permeability of these *ion channels*, making the magnificent display of different kinds of MPs possible. With each sequential event of this ionic transfer between cytoplasm and ECF, a characteristic electric potential is fired across the membrane. The rapidity with which ion transfer takes place across a neuronal membrane is quite remarkable – a single ion channel can conduct about a million ions per second! This reflects the rapidity with which signals pass through the nerves – some nerves have a conduction velocity of up to *120 meters per second*! It is no wonder that you are able to swat at a mosquito within a millisecond of noticing it on your arm.

Naturally, as time passes, this continuous transfer of ions causes a state of equilibrium (where no further ionic transfer takes place) – hence, *a polarized state has to be restored* by the action of Na^+/K^+ pump, as described above.

Now we will briefly study the common types of electrical potentials that occur in neurons and how they are generated. When excited the neuronal membrane opens up the ion channels to allow ions to pass through. However, they open only for a transient period on excitation, and close

TABLE 2.1 The Overall Effect of Ionic Transfer

Channel excitation	Ionic shift	Effect	Result	Signal pass
Sodium channel	Na^+ influx	Depolarization	Excitation	Allowed
Potassium channel	K^+ efflux	Hyperpolarization	Inhibition	Stopped
Chloride channel	Cl^- influx	Hyperpolarization	Inhibition	Stopped

down promptly. *Sodium channels* cause *influx* (entry to inside) of Na^+ ions into the neurons which causes the negative charge inside it to *decrease* (because of the relative *increase* of positive charges inside) – this is called ***depolarization*** ('de-' signifies reducing or reversing polarity; Fig. 2.8). In a state of depolarization the neurons are excited and the signals between the neurons or across the nerves are allowed to pass (Table 2.1).

On the other hand, *potassium channels* cause *efflux* (exit to outside) of K^+ ions, resulting in an *increase* of negative charge inside the neuron (because of the relative *decrease* of positive charges inside) – this is called ***hyperpolarization*** ('hyper' signifies increased polarity). In a state of hyperpolarization the signals between the neurons or across the nerves are not passed, which means that the message is blocked (Table 2.1). Or sometimes, the ***chloride channels*** cause an influx of Cl^- ions (increasing negative charges inside) into the neuron, which also results in hyperpolarization (see Fig. 2.9). ***Calcium channels*** also influence excitation indirectly – a *decrease* in Ca^{2+} concentration in ECF *increases* excitability of neurons and muscles by decreasing the amount of depolarization necessary for excitation; whereas increased Ca^{2+} levels in ECF *'stabilizes the membrane'* by decreasing excitability.

The *Goldman equation* demonstrates that at any given time the voltage of MPs depends on the *rate* of outward movement of K^+ ions relative to the rate of inward shift of Na^+ ions (Hardin et al. 2012, pp. 370–371). The *Nernst equation* states that for every 10-fold increase in cation gradient, the voltage of MP changes by −58 mV. This means that the strength of voltage potential depends on the concentration of ions (Hardin et al. 2012, p. 369).

Types of Ion Channels

It is clear from the above discussion that for MPs to be generated, first the ion channels must open. There are at least three ways by which ion channels can be triggered to open:

1. Some ion channels open up when chemicals bind to them. These chemicals are called ligands, hence these channels are called ***ligand-gated ion channels*** (see Figs 2.5 & 7.3). The ligands involved in the CNS are called *neurotransmitters* (NT). Ligand-gated channels are chiefly present on dendrites. Some types of Na^+, K^+ and Cl^- ion

channels are examples of these channels. They are responsible for the generation of ***graded potentials*** (see below). This is further discussed in Chapter 7.

2. Electrically induced ion channels are called ***voltage-gated channels*** (see Fig. 2.5). They depend on voltage fluctuations across the membrane for their opening. Some types of Na^+, K^+ and Ca^{++} ion channels are voltage-gated channels. They are responsible for the production of ***APs*** in axons (see below).
3. There are also some ***mechanosensitive channels,*** which respond to mechanical forces applied to the membrane, as seen in touch and pressure receptors in the skin and joints (Hardin et al., p. 205).

Action Potentials

APs (Fig. 2.3) are large electric potentials generated by the neuronal membranes upon their excitation. Whereas MPs are small in voltage and occur continuously in the background of a neuron, APs are large in voltage and appear only when a neuron is excited. Thus, the resting phase of a neuron is represented by MPs, whereas APs represent its excitation phase (i.e. when signals pass a neuron). APs are generated by the interplay of *voltage-gated* Na^+ and K^+ *channels* present in the axon hillock. The rapid and synchronized excitation of these channels results in the generation of APs (see below).

APs can only be generated when the neuron is sufficiently excited electrically. This means that many small signals must pass to the neuron to generate APs. When sub-threshold voltage is applied, these voltage-gated channels fail to open at all and no APs are fired. This confers them a unique property, elegantly called ***all-or-none phenomenon,*** implying that an AP can either be generated *fully* or *not at all* – and there is no half-way.

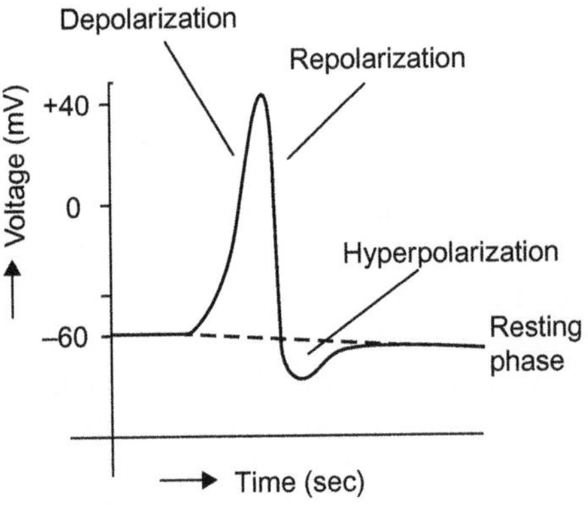

FIGURE 2.3 Action potential.

Once APs are generated at the axon hillock, they propagate along the axon until they reach the terminal bulb, where contact is made with another neuron to which they pass the electrical signal (Fig. 2.8). All nerves in the body are bunches of axons carrying such signals, which are finally transferred to muscle cells or glands to have their effect (see above, Signals that pass down). APs are self-propagating and are conducted spontaneously along the entire axon length at speeds ranging from ***4 to 120 meters per second*** (depending on the size and myelination of the axon, see below). Their amplitude is characteristically at ***100 mV***. Another important characteristic of APs is that they can propagate along the axons for any distance without losing strength because of the characteristic distribution of ion channels in the axon (Hardin et al. 2012, p. 378).

How Are APs Generated?

The generation of AP can be explained entirely on the basis of rapid Na^+ and K^+ ionic movement across the membrane. The mechanism of its generation may be described in 3 phases (Fig. 2.3).

1. *The **depolarizing phase***: all the sodium channels open simultaneously to allow an inward rush of Na^+ into the cell, and the electrical charge inside the neuron reaches to about +40 mV. This produces the ascending phase of AP.
2. *The **repolarization phase***: once the peak excitation is reached, the repolarization phase starts by two mechanisms:
   - **a.** By the inactivation of sodium channels, and
   - **b.** By the activation of potassium channels, which drives K^+ out of the neuron, lessening the positive charge inside. This produces the descending phase of AP.
3. *The **hyperpolarization phase***: at the end of an AP, most neurons show a transient hyperpolarization. This happens because of continued efflux of K^+ ions making the inside charge even more negative. This produces the characteristic 'undershoot' of AP (Fig. 2.3).

Graded Potentials

In contrast to APs, graded potentials are smaller voltage potentials which vary in their voltage between **1** and **50 mV** (hence called 'graded' potentials). They are generated by the interplay of *ligand-gated channels* which are located chiefly in the membranes of dendrites and a few in the somata (Figs 2.4 & 2.5). These ligand-gated channels of the dendrites are excited by the NTs, which are released at the terminal bulb of axon. These small and variable potentials pass along the dendrites and soma, and finally collect at the axon hillock to generate APs. If the sum of these graded potentials is of sufficient voltage, then only an AP is generated at the axon hillock; if their collective voltage

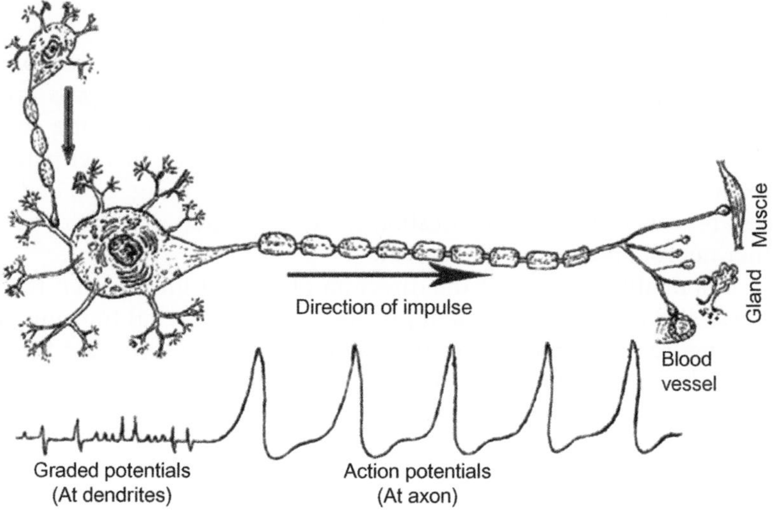

FIGURE 2.4 Parts of neuron and types of membrane potentials.

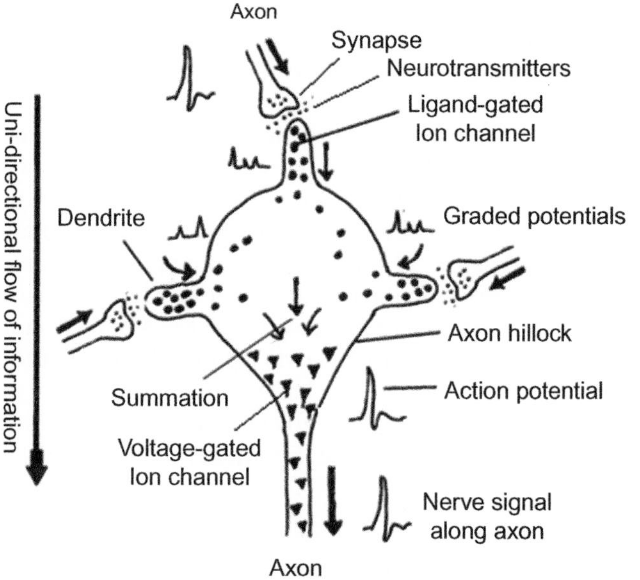

FIGURE 2.5 Schematic diagram of neuron showing ion channels and electrical potentials.

is less than the threshold value, then no AP is generated (see all-or-none phenomenon above). These APs thus generated propagate along the axon as nerve impulses, which are then conducted further down along the axon and pass to the next neuron (Signaling activity of neurons: a summary, below).

Unidirectional Flow of Information

It can be seen from the above discussion that there is a direction of signal passage in neurons. The axons carry the signals and pass on to the dendrites, and from the dendrites the signals pass over the soma, and from there they collect again at the axon hillock, and further pass on the axon – and from here the next neuron gets excited (Fig. 2.5), and the process repeats (until the final destination of the signal is reached – be it another neuron, muscle or gland etc.).

All this finally means that the direction of flow of information is *unidirectional,* i.e. there is an inherent *unidirectionality* in signal transmission across a neuron – a signal passes from a dendrite to the soma and then to the axon, but not in the opposite direction. This is because the synaptic transmission can permit conduction in one direction only.

SOMA

The body of a neuron, called the *soma,* like any other cell, contains all the organelles – nucleus with nucleolus, mitochondria, ribosomes, Golgi apparatus, endoplasmic reticulum (*Nissl bodies*), lysosomes and other necessary paraphernalia – needed to run a cell's routine administration (see Fig. 1.6).

The abundance of ribosomal aggregates (protein manufacturing units) in its cytoplasm reflects the huge turnover of proteins and enzymes needed for signal generation and transmission. The DNA present in its nucleus synthesizes the specific proteins and NTs that are used in signaling mechanism, and the synthesized NTs are loaded into small membrane-bound sacs, called *synaptic vesicles,* which are transported along the axon to reach its terminal bulb where they are released into the synaptic junction (see Fig. 2.7). The abundance of mitochondria ('power-houses' of the cell), on the other hand, denotes the high level of energy consumed by the neuron to perform its duties. The brain, though small in weight, gobbles up huge amounts of glucose and oxygen (see 'Energetics', below).

Cytoskeleton

The soma contains numerous protein strands running across its width and breadth, which not only provide its structural integrity but also help in transportation of synaptic vesicles from the soma to the axon's terminal bulb. Cytoskeleton has three important components: microtubules, microfilaments and intermediate filaments. Cytoskeleton is present not only in the soma, but extends into the entire length of the axon and into the dendrites. At the terminal bulb of the axon, the cytoskeleton anchors the vesicles until they are required to be released into the synaptic cleft (see Fig. 2.7). A layer of cytoskeleton also coats the inner surface of the neuronal membrane, and is thought to be responsible for holding the membrane proteins in place. It has been shown experimentally that if cytoskeleton is disrupted the function of membrane becomes faulty. Apart from these functions, it appears that cytoskeleton is involved in a variety of other functions, and this is a hot topic among many contemporary neuroscientists.

DENDRITES

Dendrites are about 2 μm in length, approximately 5–7 in number, and they project directly from the soma and branch extensively. They usually form tree-like arborization around the neuron, called ***dendritic tree*** (Fig. 2.6). Dendrites contain numerous ribosomes, smooth endoplasmic reticulum, Golgi apparatus and cytoskeletal structures, which show that there is a high degree of protein synthesizing activity in the dendrites during signal transmission (see Ch. 6, p. 115).

It is shown that dendrites have extensive connections with the axons in the form of axodendritic synapses, which form an important mode of communication between neurons (see Synapse below and Ch. 6, p. 110).

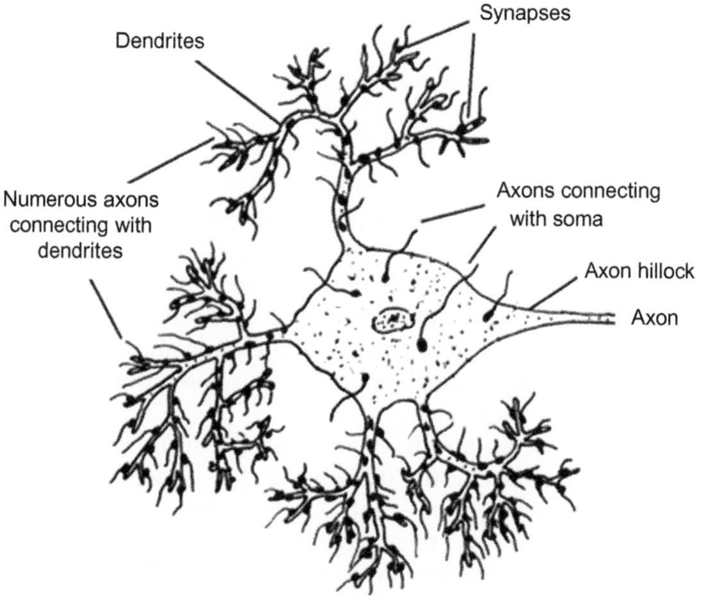

FIGURE 2.6 Dendritic tree with many synapses.

They are the chief sensors of a neuron, in the sense that the dendrites receive the incoming signals first. It has been estimated that about 75% of the dendritic membrane participate in synaptic transmission (Fig. 2.6). Dendrites are highly specialized in their function; see Chapter 6 for further discussion.

AXON

Each neuron gives out a single axon, which takes origin from a specialized funnel-shaped region called axon hillock (Fig. 2.4). The axon may extend to varying lengths – spinal motor neurons, for instance, give out axons measuring a meter in length to reach out for the legs, whereas most interneurons have very short axons to connect with the neighboring neurons. The terminal portion of the axon shows an expanded structure called *terminal bulb*, which participates in synapse formation.

There are several functional differences between dendrites and axons. Axons characteristically contain voltage-gated ion channels in their membranes (Fig. 2.5), which respond only to voltage fluctuations, and this forms the basis for the generation of APs, as described above. Consequently, the axons have a propensity to transmit only uniform APs, and cannot transmit graded potentials (Fig. 2.5), whereas dendrites are rich in ligand-gated ion channels; consequently dendrites are specialists in generating and transmitting graded potentials. The differences between axons and dendrites are presented in Table 2.2; however, an elaborate analysis of these differences is undertaken in Chapter 6.

Myelinated Axons

The axons of many neurons are myelinated. Myelin is a protein–lipid complex secreted by oligodendrocytes, which wraps around and insulates the axons. Generally, axons of more than 2 μm in length are myelinated. Axons show gaps in myelination throughout their length called ***nodes of Ranvier*** (Fig. 2.4). Myelination insulates the axon thus preventing dissipation of ionic energy during signal transmission – this makes a signal travel fast across the axon.

The conduction of an impulse over long distances is facilitated in a myelinated axon because of the tremendous flux of ions at the nodes of Ranvier. Thus, the APs generated at a node of Ranvier 'jump' from one node to another because of the high resistance and low capacitance at myelinated portions. This jumping transmission of nerve impulse across myelinated axons is called *saltatory conduction*, and signal transmission this way can be very rapid (up to 120 m/s). The number of Na^+ channels at

TABLE 2.2 Differences between Axons and Dendrites

	Feature	Axon	Dendrite
1	Number and length	Single – long	Many – short
2	Signal direction	Takes signal from soma to axon	Takes signal from dendrite to soma
3	Function	Sending signals	Receiving signals
4	Ion channels	Voltage-gated	Ligand-gated
5	Membrane potentials (MPs)	Action potentials (APs)	Graded potentials
6	Voltage	Uniform 100 mV	Graded 1–50 mV
7	Variability of MPs	Fixed – all-or-none	Graded – variable
8	Distance they can travel	Long distances	Short distances
9	Ribosomes for protein synthesis	Few	Numerous
10	Neurofilaments	Inactive phase (highly phosphorylated)	Active phase (poorly phosphorylated)

the nodes of Ranvier are between 2000 and 12 000, whereas at internodal segments there are about 25 (Barrett et al. 2012, p. 91.) – this facilitates saltatory conduction.

In the ***unmyelinated axons*** the Na^+ channels are far less in number and are uniformly distributed, hence the conduction is slower (about 4 milliseconds).

SYNAPSE

The neurons are connected to each other by way of synaptic junctions, and the whole study of the brain's functions is centered on these synapses. Neurons communicate with each other in two ways – ***chemical synapses*** and ***electrical synapses*** (and a third, ill-understood, type called *ephaptic coupling*; Ch. 8, p. 152). Chemical synapses are very common in the CNS; although a little bit slower in conduction (2–4 ms) they are *superbly versatile* and are designed to *respond variably* to a given stimulus (which is not possible in electrical synapses). The chemicals which mediate synaptic transmission are called neurotransmitters, and are secreted into the synapse by the axon.

In contrast, electrical synaptic conduction is much faster (0.2 ms) and occurs *rarely* in the CNS, but is seen commonly in heart muscle – but we will focus on these issues in Chapter 8, where it is relevant.

Note: A synapse in this book generally means only a *chemical synapse*, unless specified.

Types of Synapses

- ***Axodendritic synapse***: this typically forms when the membrane of the terminal bulb of an axon comes into contact with that of a dendrite (Fig. 2.8). The axodendritic synapses are either excitatory or inhibitory. Axodendritic synapses are the commonest type of synapse in the CNS and are described fully in Chapter 6, p. 110.
- ***Axosomatic synapse***: here the axon synapses directly with the soma – these may be excitatory or inhibitory.
- ***Axoaxonic synapse***: here the axon connects with axon of another neuron – these are usually inhibitory, as they make the receiving neuron less excitable.
- ***Dendro-dendritic***: synapse between dendrite and dendrite (Ch. 8, p. 149).
- ***Dendro-axonic synapse***: a dendrite communicates with axon (rare).
- Synapse between soma and another soma.
- ***Neuromuscular junction***: an axon may finally end on muscle fibers (commonest synapse in the peripheral nervous system).

Synapse in the following description means an axodendritic synapse, because they are the commonest synapse in the CNS. In a synapse, the terminal bulb's membrane forms the ***presynaptic membrane*** (Fig. 2.7) and the dendritic membrane of the receiving neuron forms the ***postsynaptic membrane***, and the potential gap between the two is the ***synaptic cleft*** (usually about 20–30nm). The presynaptic membrane (i.e. axon's membrane)

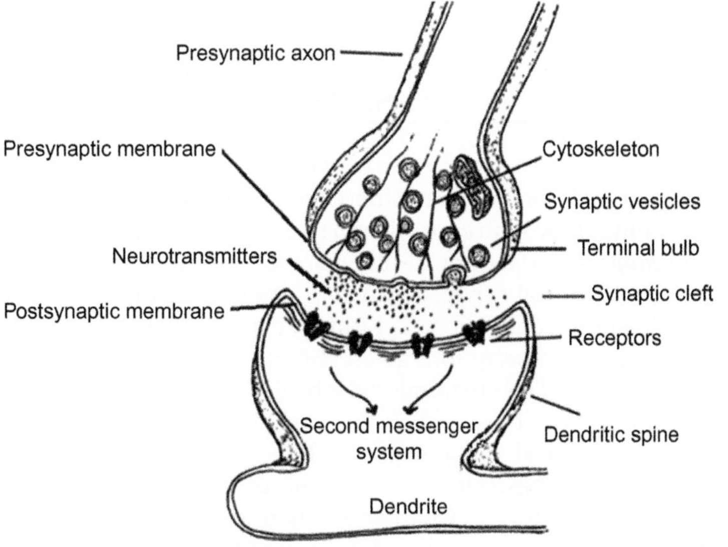

FIGURE 2.7 Axodendritic synapse.

secretes NTs into the synaptic cleft, and the NT acts on the postsynaptic membrane to show the desired effect. The following section describes more about the electrical activity at the synaptic junctions.

The most important function of a chemical synapse is its ability to show ***synaptic plasticity***, and this is the fundamental property of neurons that confers the human brain its capacity for memory and learning, and intelligence – which in turn forms the basis of all higher intellectual functions. This special feature is described in Chapter 3, pp. 67–71.

SIGNALING ACTIVITY OF NEURONS: A SUMMARY

Now we will collate all the above knowledge and summarize how signals pass across the neurons. Most of the signaling activity in the neurons is ***electrochemical*** in nature, which means that the neurons receive various chemicals in the form of NTs and generate electrical potentials. These electrical potentials travel along the axons as nerve impulses and excite other neurons again by releasing NTs at their synapses. Thus, synapses are where electrical energy is *transduced* (converted) into chemical energy and back into electrical energy (Fig. 2.8), as we will now see. This is called ***signal transduction*** (Ch. 7, p. 129). Below, we will describe the signaling activity at a typical axodendritic synapse.

Signaling activity begins with the arrival of APs at the ***presynaptic axonic membrane***, which causes opening of voltage-sensitive *calcium channels* present in the terminal bulb membrane. This electrical excitation results in a sudden influx of Ca^{2+} ions into the terminal bulb, which in turn activates an enzyme system (involving *protein kinases*). These events uncouple (=release) the vesicles containing NT from the strands of cytoskeleton in the terminal bulb, like berries falling from twigs. The released vesicles reach the presynaptic membrane to 'dock and fuse' with it (Fig. 2.7), and

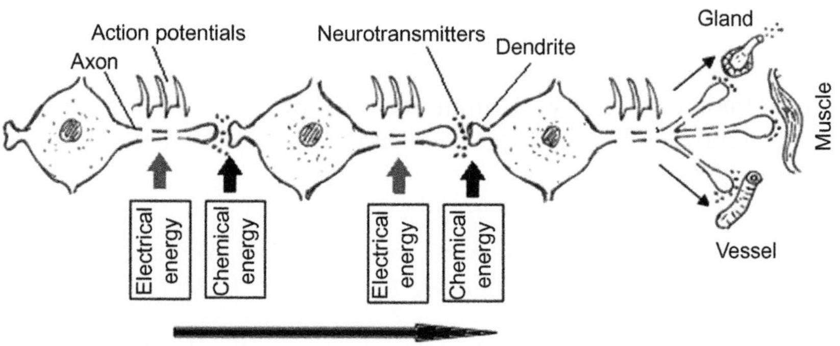

FIGURE 2.8 Electrochemical transmission (signal transduction).

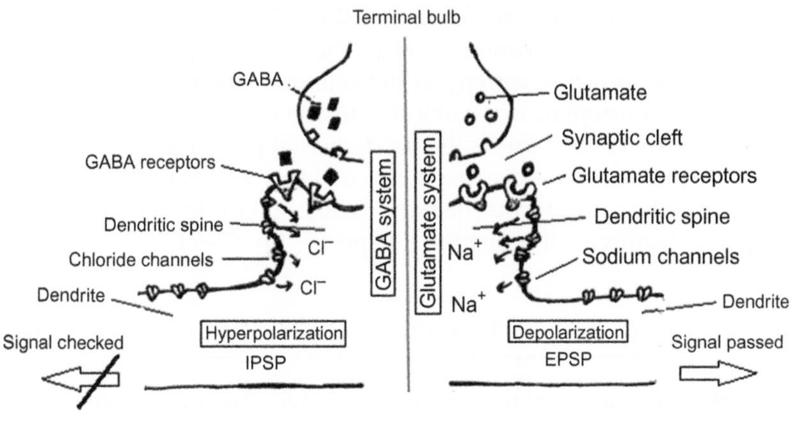

FIGURE 2.9 Neurotransmitters and EPSP – IPSP.

open up to spill their contents into the synaptic cleft. All these events occur with great rapidity.

Postsynaptic dendritic membrane contains the receptors for the NTs released into the synaptic cleft. The NTs bind to these receptors, which causes ionic disturbances at the membrane. This ionic transfer across the postsynaptic membrane triggers small MPs called *postsynaptic potentials* (Hardin et al. 2012, p. 388). Usually the effect of NTs is very rapid but transient (short-lived), because the released NTs are quickly destroyed by the enzymes in the synaptic cleft, or the NTs are taken back into the presynaptic terminal (reuptake), or the NTs are absorbed by the surrounding neuroglia (Ch. 1, p. 18). There is growing evidence that the glial system controls the electrical activity of neurons in this way, to some extent (Han et al. 2012).

There are chiefly two kinds of postsynaptic potentials: 1) those that allow the signal to pass, called *excitatory postsynaptic potentials* (EPSP), and the *facilitatory* NTs that cause them; 2) those that keep in check the signals are called *inhibitory postsynaptic potentials* (IPSP) and the NTs that cause them are *inhibitory* NTs. Facilitatory NTs cause an influx of Na^+ or Ca^{2+} that causes the membrane to undergo depolarization (see 'Polarization, depolarization and hyperpolarization' above) – and the signal is allowed to pass. Inhibitory NTs cause an influx of Cl^- or efflux of K^+ that causes the membrane to undergo hyperpolarization (see also above) – the signal is checked (Fig. 2.9).

Look at the following sequence of events:

→ Excitatory NT (e.g., glutamate) secreted by the axon terminal into the synaptic cleft (the glutamate receptors are further discussed in the Ch. 3, p. 68)

→ Glutamate acts on the glutamate receptors in the dendritic membrane

→ Glutamate receptors excite the sodium channels
→ Sodium channels allow Na^+ ions into the dendrite
→ Na^+ ions causes *depolarization* of the dendritic membrane
→ Depolarization causes *excitation* of the dendritic membrane
→ These signals are **EPSP**
→ Finally, the ***signal is allowed*** to pass.

Or, look at the following sequence of events:

→ Inhibitory NT (e.g., GABA) secreted by the axon terminal into the synaptic cleft
→ GABA acts on GABA receptors in the dendritic membrane
→ GABA receptors excite the chloride channels
→ Chloride channels allow Cl^- ions into the dendrite
→ Cl^- ions causes *hyperpolarization* of the dendritic membrane
→ Hyperpolarization causes *inhibition* of the dendrite membrane
→ These signals are **IPSP**
→ Finally, ***signal is checked*** from passing.

The EPSPs thus generated at the dendrites are of varying strengths and are thus called ***graded potentials*** (see p. 40). These graded potentials are passed on to the soma where they join with other graded potentials coming down from other dendrites, and finally they join with each other. This process of adding up several EPSP is called ***summation*** (Fig. 2.5). If their collective strength is sufficient to excite the axon hillock then the axon hillock fires an AP. If their collected strength is not sufficient they will not have any effect.

However, it must be noted that some of the dendrites generate IPSPs, as discussed above, and these will cause membrane hyperpolarization and actually weaken the summation. The EPSP and IPSP (Fig. 2.8) are in fact small localized ionic events in the dendritic tree, and the ultimate response of a neuron depends on the algebraic sum of several of these small EPSPs and IPSPs generated at the dendritic tree (Barrett et al. 2012, pp. 124–126); if the sum of excitatory signals exceeds the inhibitory signals then an AP is fired at the axon hillock, or *vice versa*. It is interesting to note that many sedative drugs (like Valium®) act by opening up the Cl^- channels and generating more IPSPs at the dendrites and suppressing excitatory signals.

These series of events continue down the chains of interlinked neurons until the signal passes to their final destination (Figs 2.1 & 2.8).

NEUROTRANSMITTERS

A neurotransmitter (NT) is a small molecule released at the presynaptic axonal membrane of one neuron into the synaptic cleft to bind with the receptors present in the postsynaptic membrane of another neuron (or muscle fiber), which results in either excitation or inhibition of the passage of signals across the synapse. The receptor is the lock and NT is the key – the key turns the lock to either 'off' or 'on'. A neurotransmitter is the currency with which neurons talk to each other, and consequently NTs are concerned with all phenomena related to the brain, such as memory, learning, intelligence etc.

An excitatory NT causes depolarization, whereas inhibitory NT causes hyperpolarization, as discussed above. All these NTs are synthesized in the somata (bodies of neuron) and are stored in the *neurosecretory vesicles* (Fig. 2.7), which are then transported along the axon to reach synaptic junction. Current research in neuroscience has identified a class of molecules, called ***neuromodulators,*** which indirectly influence the activity of NTs in signal transmission, thus modulating the neuronal response.

A Few Examples of NTs

1. Acetylcholine (ACh): a common NT in the CNS, and the most extensively studied. ACh is an excitatory NT which acts by increasing Na^+ permeability. The neurons that use ACh are called *cholinergic neurons*. ACh is the chief NT of motor neurons, preganglionic autonomic neurons and postganglionic parasympathetic neurons. It is also the most common NT outside the CNS, for example in muscles (at neuromuscular junctions). ACh is deactivated rapidly in the synaptic cleft by an enzyme called *cholinesterase.*
2. Catecholamines: a class of molecules derived from amino acid *tyrosine* – the common examples are *adrenaline, noradrenaline* and *dopamine*. The synapses involved are called 'adrenergic synapses' which are present throughout CNS and PNS. They are excitatory.
   a. ***Adrenaline and noradrenaline***: these are essential synaptic transmitters which mediate the *'fight-flight-fright'* response by the body. The effect of adrenaline is to improve the activity of heart and muscles so that they can work better and faster. Adrenaline also breaks down glycogen stores to supply glucose to the exercising muscles. They act by binding to *adrenergic receptors,* which are G protein-linked receptors (Ch. 7, p. 131).
   b. ***Dopamine***: dopaminergic neurons are present extensively in the cortex, diencephalon and midbrain, especially concentrated in areas such as substantia nigra, corpus striatum and limbic system. Dopamine is excitatory (Parkinson's disease, Ch. 1, p. 25).
3. Serotonin (5-hydroxytryptamine or 5-HT): serotonin is a monoamine compound. Serotonergic neurons are predominantly distributed in the brainstem, the axons of which extensively ramify throughout the brain and spinal cord. They are also seen in the gastrointestinal

system. Serotonin is excitatory, and works by closing K^+ channels. Dietary intake of tryptophan-rich food (meat, eggs, beans etc.) can increase serotonin content of the brain – this may have some role in regulating appetite, sleep patterns, and mood. It gives a sense of well-being. Serotonergic system in the brain appears to regulate impulse control and mood, and there is some evidence that this system is seriously impaired by alcohol.

4. Histamine: especially seen in posterior hypothalamus and their axons connect to all areas in the CNS. Histamine appears to be involved in most hypothalamic functions.
5. GABA (γ-aminobutyric acid): the major inhibitory NT extensively distributed throughout the CNS. GABA acts by opening Cl^- channels thereby hyperpolarizing the postsynaptic membrane. Diazepam (Valium®) shows its tranquilizing effect by potentiating the action of GABA receptors. Alcohol and drugs for general anesthesia appear to act via the GABA system. It is interesting to note that GABA's overall inhibitory role in the CNS is responsible for 'cutting-down' unnecessary sensations in the background – enabling human beings to perceive important signals more attentively (Barrett et al., 2012, p. 142).
6. Glutamate and aspartate: these are the major excitatory NTs distributed wide across the CNS. Glutamate is discussed in Chapter 3, p. 68, as it is concerned with memory.
7. Glycine: an inhibitory NT presents in the brainstem and spinal cord, and is concerned with local circuits.
8. Nitric oxide (NO): a gas which plays a very important role in the autonomic and enteric nervous system. NO controls blood supply to various organs by mediating smooth muscle relaxation. *Sildenafil* (Viagra®) is a drug used in penile erectile dysfunction (impotence) which acts by potentiating the effects of NO. It also has a role in long-term memory.
9. Neuropeptides: there more than 50 distributed abundantly throughout the nervous system. These short-chain amino acids are produced by the cleavage of precursor proteins. They generally co-exist with other NTs. Their effect is generally long-lasting. A few examples: ***Substance P*** is the prototype neuropeptide, which is the major NT in the brain and spinal cord. Its chief action is prolonged excitation of nociceptive (pain-carrying) centers. This also triggers inflammation and other myriad effects after injury. ***Vasoactive intestinal polypeptide*** (VIP) is an excitatory neuropeptide present in the CNS. ***Somatostatin*** is widely distributed in the CNS. ***Encephalins and β-endorphins*** belong to this group. Encephalins are involved in the pain perception. Painkillers like morphine act by blocking the receptors for encephalin.

10. ***Anandamide:*** derived from arachidonic acid and found in hippocampus, basal ganglia and cerebellum. It is shown to have a definite analgesic effect (Barrett et al. 2012, p. 173). The term *ananda* is derived from *Sanskrit* and means bliss.
11. Others: many substances influence the function of the brain by neuromodulation either directly or indirectly e.g., prostaglandins, steroids, ATP, adenosine (caffeine is adenosine antagonist – hence the refreshing effect of coffee) are a few examples. Placebos, acupuncture and other unconventional methods appear to act on the CNS by affecting some of these systems.

A note on neurotrophins: in order to survive, neurons require not only energy, but also many other 'stimulating factors'. These factors are either produced by the surrounding astrocytes or the effectors they supply (such as muscles). These factors are called neurotrophins, which travel up and bind to the receptors present over the soma and provide a constant stimulus for the neurons to keep themselves active. This is the reason why neurons tend to atrophy (die) when the muscles they supply become nonfunctional. Nerve growth factor (NGF), brain-derived neurotrophic factor (BDNF), NT-3, NT4/5 are some examples of neurotrophins.

ENERGETICS

The human brain needs huge amounts of energy in the form of glucose (120 g per day) for its normal function and it is interesting to note that the brain can utilize no other nutrient for its energy than glucose. The brain consumes a lot of oxygen too – though the brain weighs only about 2% of the body's total weight, it uses about 20% of the body's total oxygen consumption (Hardin et al. 2012, p. 243).

Much of this energy is used to generate MPs essential for the transmission of nerve impulses. The restoration to normal polarity of membrane and the subsequent firing of MPs and APs is energetically a costly affair. It is estimated that up to two-thirds of a neuron's energy is spent in maintaining polarity by activating Na^+/K^+ pump. A significant amount of energy is also consumed in vesicular transport along the axons. No wonder most of us have found that mental activity can be really exhausting.

In these first two chapters, we have learned about the anatomy and physiology of the brain; now we will proceed with the study of human memory in Chapter 3, which is the fundamental property of the human brain.

CHAPTER

3

Memory

OVERVIEW

A strikingly characteristic feature of human beings (and some other higher animals) is their ability to alter behavior on the basis of experience. This is brought about by a process called learning, and the basis for this is firmly rooted in a facility called memory. Memory is the fundamental property of the nervous system and is the chief constituent of all the major mental processes. However, human memory has historically defied precise scientific definition, perhaps because of its all-pervasiveness, which confers on it a certain amount of vagueness.

The human ability to conjure up specific long-gone episodes in their personal lives is intriguing, but at the same time these long memories form a key aspect of a human being's personality – we are what we remember. Every aspect of *present* human action is the product of memory of the *past*; and the present event also bears a possible *future* speculation. In contrast, animals live mostly in the present; some higher animals are capable of going into the past to some extent, but never too far into the future. For instance, elephants in a jungle can remember the source of water for some time but will never think of storing water for the future. This peculiar human ability to think of the future (and to hoard *for* the future) has been the cause of all miseries. On the other hand, this capability of imagination is also fundamentally responsible for human civilization, as it were.

The seat of memory is the brain – and the neuron and the neuronal networks are its fundamental machinery. We know that most of these higher functions are processed somewhere in the nooks of the brain, but we have not yet precisely located any specific area in the brain for memory storage and intellectual attainment. In this chapter we discuss some of the essential physiological aspects of memory and its related phenomena.

The Biology of Thought.
DOI: http://dx.doi.org/10.1016/B978-0-12-800900-0.00003-8

HOW IS MEMORY INITIATED?

Various ***stimuli*** from the external world reach the receptors, where they are converted into ***sensations*** and these sensations are sent up to the neurons in the brain where they are converted into ***perceptions*** (or ***thoughts***). Brain consolidates these thoughts into ***memory*** (Fig. 3.1). Thus, input of sensations is the first basic step in the formation of memory, without which no memory is possible. For example, when you look at a flower, such as a rose, the light rays impinge on the retina and form *sensations* of light, which are then sent to the neurons in the brain where they are perceived as *perceptions* (or *thoughts*) of light. These thoughts are reconstructed into a complex idea called 'rose' in the brain and this idea is stored as *memory*. Now we will see exactly how the peripheral sensations reach the brain by taking vision as an example.

General Mechanism of Vision

The human retina consists of photoreceptor (light-sensitive) cells called *rods* and *cones*, arranged in layers. There are about 120 million rods and 6 million cones in each human retina. Upon receiving the light, the rods and cones are excited and send signals to specialized nerve cells situated deep in the retina, called *bipolar cells* (Fig. 3.2). The bipolar cells in turn relay these signals to another group of neurons called the *ganglion cells*. Finally, the ganglion cells give out nerves which bundle up to form the *optic nerve* (the chief nerve of the eyeball). There are approximately 1.2 million axons in the optic nerve. Thus, more than 120 million photosensitive cells converge on a few bipolar cells, the axons of which converge on still fewer ganglion cells, ultimately giving out only 1.2 million axons in each optic nerve. This means that there is a good deal of *convergence* in the transmission of messages from retina to the optic nerve (Fig. 3.2). The rods and cones are connected by special intermediary cells called the *horizontal cells*, which assist in modifying the signals, whereas the bipolar cells are interconnected by another set of intermediary cells called the *amacrine cells*, which can modify the signals before they reach the ganglion cells.

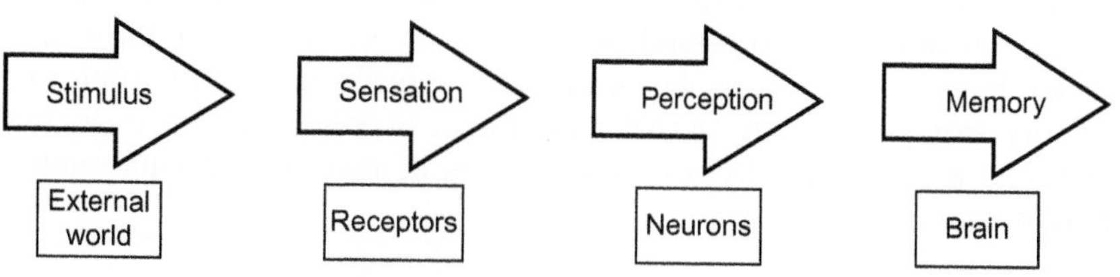

FIGURE 3.1 Steps of formation of memory.

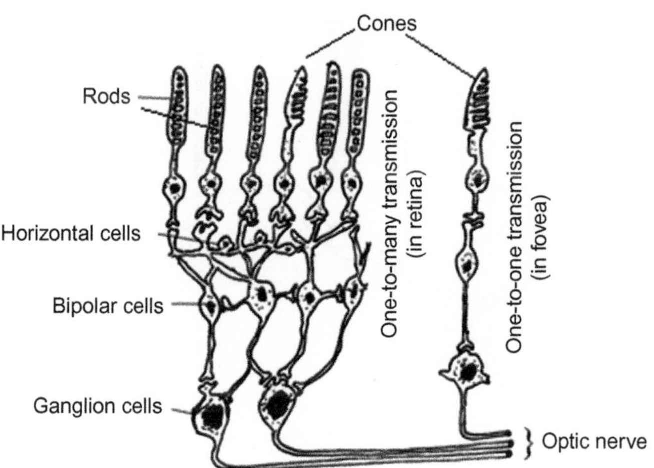

FIGURE 3.2 Photoreceptor cells of retina.

The rods are extremely sensitive to light and are responsible for vision in dim light, for instance at night (*scotopic vision*), but they are not capable of responding to the color of the object. The cones are responsive only to bright light (*photopic vision*) and they are responsible for perceiving minute details of the object and identification of colors.

In the human eye, cones are concentrated in a central portion of the retina called *fovea centralis*, where the cones are densely packed, and this area contains virtually no rods. The way that cones in the fovea send signals to the brain is unique – each cone in this area is connected to a single bipolar cell, and each bipolar cell is connected to a single ganglion cell (Fig. 3.2; Barrett et al. 2012, p. 180). This *one-to-one transmission* provides these cones with a direct pathway to the visual cortex (without convergence) thus increasing visual acuity (whereas in the rest of the retina nearly 100 rods and cones share one bipolar and one ganglion cell, hence the visual acuity is far less than in the fovea). That is the reason we tend to narrow our eyes and fix our gaze to see an object clearly – this allows the image to fall right upon the fovea, enhancing the acuity to its greatest extent.

Visual Neural Pathways

Optic nerves on both sides meet at the *optic chiasma* (Fig. 3.3) where axons from both sides cross over, which gives us the phenomenon of *binocular vision*. The bundles of axons that leave the optic chiasma are called the *optic tracts*. These axons ultimately end on the neurons located in a structure called the ***lateral geniculate body***, which is a part of the thalamus. There is a good deal of *divergence* from this point onwards – which

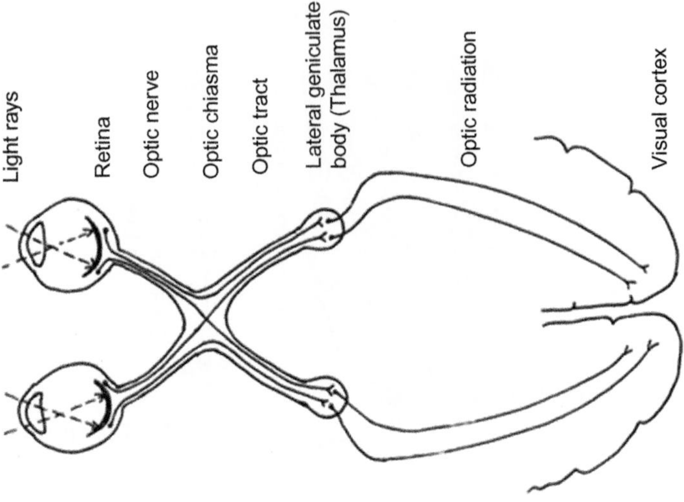

FIGURE 3.3 Visual neural pathways.

means that there are 1000 times more axons from the lateral geniculate body than in the optic nerve, and this not only serves the purpose of vision but also the coordination of eyeball movements and various reflexes of the eye. Most of the axons from the neurons of lateral geniculate body project to the visual cortex, but some project to the midbrain and some to thalamus and other cortical regions; for example, if bright light is shone into *one* eye not only is the light sensation perceived but *both* pupils constrict – and this is taken care of by these additional connections. However, a perfect spatial arrangement is maintained throughout the transmission of the image from retina up to visual cortex to avoid distortion.

To summarize, the *first image* is formed in the photoreceptor cells (rods and cones); the *second image* is formed in the bipolar cells; finally, a *third image* is formed in the ganglion cells. There may be some modification in the first and second images but it appears that there is little change in the third image, and it passes unchanged to the visual cortex (Barrett et al. 2012, p. 185). The perception of image in the cortex is discussed in Chapter 5, pp. 106–108, where it is relevant.

Other Sensations

All peripheral sensations, such as smell, taste, sound, touch, pain etc., follow almost similar pathways by passing through various filters such as the spinal cord or midbrain and thalamus. The perception of general senses such as touch and pain will be discussed in Chapter 5, pp. 94–98;

the pathways of the other sensory modalities are not discussed as this is outside the scope of this book. Now we will see how memory is formed.

FORMATION OF MEMORY

Human memory is chiefly divided into three categories, depending on the duration of storage (see also Ch. 8).

- Sensory memory
- Short-term memory (STM)
- Long-term memory (LTM).

Sensory Memory

The sensations from the external world are perceived by the cortical areas involved, such as visual sensations by the visual cortex, sounds by the auditory cortex, touch by the somatosensory cortex, etc. The sensations, as they reach the brain, are transiently registered in the form of *sensory memory* and this memory is retained for only a few hundred milliseconds to 1–2 seconds before it becomes short-term memory (STM). In fact, this is the first step in the process of perception. All sensory memory does not pass on to STM (see example below) – only when sensory memory enjoys conscious appreciation does it become short-term memory (Fig. 3.4).

For example, when we look at a beautiful scene of wonderful blue mountains with an elegant azure lake at their foot and with flowery bushes to one side, we may fail to take notice of a small dinghy (boat)

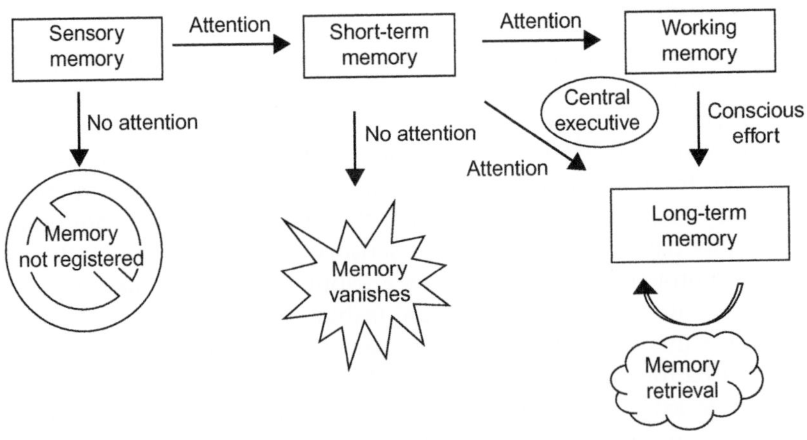

FIGURE 3.4 Progression of memory.

lying motionless in one corner of the lake. The image of the dinghy has certainly impinged on the retina, along with images of mountains, lake, bushes, etc., but we have *not registered* the boat in our memory. It can be said that the image of the dinghy gave us sensory memory – although the image of the dinghy has reached the brain along with images of other details, the dinghy was not perceived because it did not trigger conscious perception in the brain. In other words it is only a sensation, and is not stored as memory. The result is that the sensory memory thus formed may pass unnoticed by the observer. We receive so many sensory inputs day-in and day-out that we take notice of relatively few of them.

Short-Term Memory

Sensory memory thus formed may end up in two ways: it can be registered as STM; or it may be lost from consciousness forever (Fig. 3.4). The STM thus formed may also end up in two ways it may pass into long-term memory; or may vanish totally. For example, when we look at a number in the telephone directory before starting to dial, the number is temporarily stored for a few seconds and after completing the task it will vanish from our memory (unless a conscious effort is made to remember it, then it becomes LTM). The duration of STM storage ranges between 10 and 12 seconds. Studies have shown that a human being has the capacity to remember about 7±2 items at a given moment (*'the magical number seven'*). It is common knowledge that the fate of STM depends on the attention the perceiver pays – if no attention is paid the memory decays; if conscious effort is made it forms LTM. STM is acclaimed to be processed in the ***prefrontal cortex*** (PFC) (see Ch. 1, p. 11).

Working Memory

When a person needs to perform a complex task (multitasking), he requires a little more memory storage for a little longer duration and in a better organized fashion for efficient execution of the task (Fig. 3.4). For example, a student listening to his tutor and at the same time copying notes from the blackboard is *multitasking* – he has to *listen* to the tutor, *look* at the blackboard and *write* down the notes – all at the same time. His memory has to deal with various sensory signals and coordinate sensory and motor skills. These tasks are attributed to *working memory*. Here perceptions are stored in a sequential fashion in order to execute a task efficiently – this ability makes human beings skilled workers.

In 1974 Baddeley and Hitch proposed a model of working memory, in which there exists a ***central executive*** that monitors and coordinates all the sensory input. The sensory input is subdivided into two systems – ***phonological loop*** (deals with auditory, spoken and written information) and ***visuospatial sketchpad*** (deals with visual information and the spatial relation between objects). The central executive is the boss and the other two are considered subordinate systems. The phonological loop activates certain areas in the left cerebral hemisphere that are associated with the production of language, such as Wernicke's area and Broca's area (Ch. 1, pp. 11 & 12). The visuospatial information is processed in the visual cortex.

Studies have shown that when a person is asked to remember a complex problem, certain areas in the prefrontal cortex are activated (especially the dorsolateral prefrontal cortex), indicating that the central executive is located in the PFC. This central executive is the most versatile organization which appears to know how to deal with the STM and working memory – if it directs the brain to ignore the memory, it is ignored and the memory vanishes 'into thin air' (Fig. 3.4). On the other hand, if the central executive directs the brain to remember the thing or event, it can be stored for days, weeks, months or for life – and it can perhaps be recalled on demand. The real nature of this central executive is still elusive, and will be discussed in Chapter 9.

Long-Term Memory

Our student in the above example stores all the information in his working memory, but he is likely to forget things as soon as the tutorial ends. In order to retain the information he has to make a conscious effort to store it for longer periods to enable him to reproduce it in his exams (Fig. 3.4). The *duration of storage* of LTM is unlimited – it can be stored for a day, a few days, months, a few years or for a lifetime – depending on the involvement and interest of the individual concerned. Not only the duration but the *capacity of its storage* is also phenomenally large; you may even say that it is endless in a human being (some people have retained volumes of information in their lifetimes: for example, Swami Vivekananda (1863–1902) reputedly had an entire encyclopedia stored in his mind). It is not clear *where* exactly LTM is stored, but it appears that there is no specific area in the brain for its storage. However, the gross mechanism involved in the formation of LTM is fairly certain, as described below; and the exact molecular mechanism of memory is the subject of Chapter 8.

A solid LTM is a prerogative of human beings and forms the basis of all their higher intellectual functions. The moods and emotions, happiness and sadness, love and hate, and ultimately, the overall personality of an individual have deep roots in his/her LTM.

LTM is generally divided into the following two categories (Fig. 3.5):

- Explicit (declarative) memory
- Implicit (non-declarative) memory.

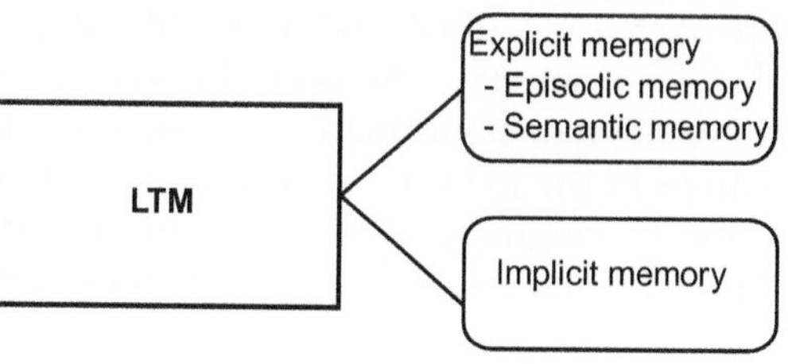

FIGURE 3.5 Types of LTM.

Explicit Memory

You may remember at the end of the day, and can perhaps describe, that day's events with some accuracy. You remember your morning breakfast; recollect an interesting piece of news from the newspaper; or the dress you wore. You remember your date of birth (perhaps your spouse's too!) for life. Students read and remember their tutorial notes, at least until their exams. All these memories reach your conscious mind and you can appreciate and describe the things you remember. This is called declarative memory or explicit memory.

In this form of memory the individual is aware of remembering and rather puts in a conscious effort to memorize, and it can be explicitly described in words. This declarative memory is fairly reliable in day-to-day life – an unfaltering student reciting her math tables, an unerring money-lender tallying his accounts, an aging grandpa pestering his grandchildren with his childhood anecdotes – and other innumerable such things stand out as examples. However, its accuracy and reliability are dependent upon various factors, such as the interest involved, the emotional settings, the time elapsed, and many such others (see 'Factors affecting memory', below).

Explicit memory is of two types – episodic and semantic memory.

Episodic Memory

Episodic memory is concerned with personal experiences (hence also called *autobiographical*) –the breakfast you had that morning, the suit you wore for last night's party – such episodic events are important for day-to-day activities. The duration of its storage largely depends on the attention an individual gives to it – you may remember the sumptuous dinner eaten at your friend's house for years, yet you may forget other details (like your friend's son's name) that are associated with the event. ***Hippocampus*** is fundamental in recording episodic memory (see 'Mechanism of LTM formation', below).

Amnesia is a condition in which a person fails to recollect episodic memory. A peculiar sort of *anterograde amnesia* called *Korsakoff's psychosis* (see 'The Papez circuit', below) occurs in some persons with chronic alcoholism (with vitamin B_1 deficiency), rendering them unable to store episodic memory – the person can remember all past memories but new memories are not recorded.

Semantic Memory

Semantic memory signifies the *meaning of words*. It is a conceptual memory and helps the individual to *assign function to a word*. The following example makes it clear. If a child is given a slate and a piece of chalk, he/she plays with them rather than writes – the child first becomes familiar with these items by forming a memory regarding the color and form of slate and chalk. The child does not know the name or purpose of slate and chalk, but gets familiar with them by their forms and shapes – this forms episodic memory. And so, as you teach the child to know them by their names, 'slate' and 'chalk', and as you teach him how to use them to write – then semantic memory begins to form. The purpose of semantic memory is to process and categorize the episodic memory into genres (see 'Alzheimer's disease', below, to better understand semantic memory). ***Medial temporal lobe*** (MTL) and ***dorsofrontal lobe*** appear to be involved in the processing of semantic memory.

Semantic memory is the most important part of human intellect – without it you may remember scores of words at length, but they all become gibberish when you do not recognize their meaning and purpose. Alexander Luria, the famous Russian neurologist, cites a person with an extraordinary mental ability to remember words or numbers for months or years, but who failed to answer simple questions from the text because he had an exceptional ability only to remember the words but not their meanings (Bruno 2002).

Alzheimer's disease is a progressive neurological disorder affecting the aging population. Those affected gradually become incapacitated by loss of long-term memory, leading to emotional imbalance and problems with language, perception and thinking, and ultimately resulting in marked *dementia* (severe loss of brain function). Though the exact cause is uncertain, neuronal damage to temporal lobe and hippocampus, and depletion of acetylcholine in the prefrontal cortex are implicated in its pathogenesis. In Alzheimer's there is a difficulty in retrieving the meaning of words or group of words from LTM. When the disease is mild the person forgets the name of his pet Alsatian dog; as the disease progresses he remembers only that it is a dog but cannot discriminate it as an Alsatian breed; and in severe forms of Alzheimer's he can only say that it is an animal but cannot say that it is a dog (Bruno 2002), because semantic memory helps to categorize the memory into genres. In later stages, the life of the affected person becomes so incapacitated by the loss of memory that he may even fail to navigate within his home.

Implicit Memory

Implicit memory is also called *non-declarative memory*, *motor memory* or *procedural memory*, and it cannot be described in words. For this memory to form, overt conscious appreciation of memory is not necessary; for example, performing skilled tasks using the hands, such as buttoning a shirt or tying a shoe lace, do not need continuous attention – they are done almost automatically. Likewise, cycling or swimming also do not depend on conscious effort, but are based on previously learned motor memories. ***Cerebellum, basal ganglia*** and ***motor cortex*** are involved in implicit memory, but of course, it is supervised by the ***cerebral cortex***.

Hippocampus, which is essential for explicit memory, is *not* needed for implicit memory. But it must be understood that for the implicit memory to form, explicit memory has to form first and train the cerebellum and other centers. Thus, for the formation of new implicit memory the presence of an intact hippocampus is a necessity; for example, in order to learn to ride a bicycle, firstly explicit memory has to form during learning, but once learned, cycling can be carried out later unconsciously, using motor memory.

MECHANISM OF LTM FORMATION

There are still many wide gaps in our knowledge regarding memory formation and storage in the human brain. However, certain aspects of LTM formation are fairly well understood, and these aspects are described as follows.

Outlines of LTM Formation

The first step in the formation of memory is the formation of sensory memory, which forms when signals reach cortical areas (visual signals to visual cortex, olfactory signals to olfactory cortex, general sensations to somatosensory cortex, etc.). For example, when you pick up a rose and smell it, the visual signals regarding its color and shape reach the visual cortex; the signals regarding touch and texture reach the somatosensory cortex; and the smell signals of it reach the olfactory cortex – all these sensations are reconstructed into a composite idea of a rose flower (Fig. 3.6). We will now see how this happens.

Hippocampus (Ch. 1, p. 13) plays a crucial role in the formation of LTM. The general mechanism is as follows: the sensory perceptions that are appreciated in the cortex form STM for a brief period, and this STM is then referred to the hippocampus (Ch. 1, p. 14) where the information is processed. It is widely believed that this processed information from the hippocampus is promptly sent back to the respective cortices to be stored

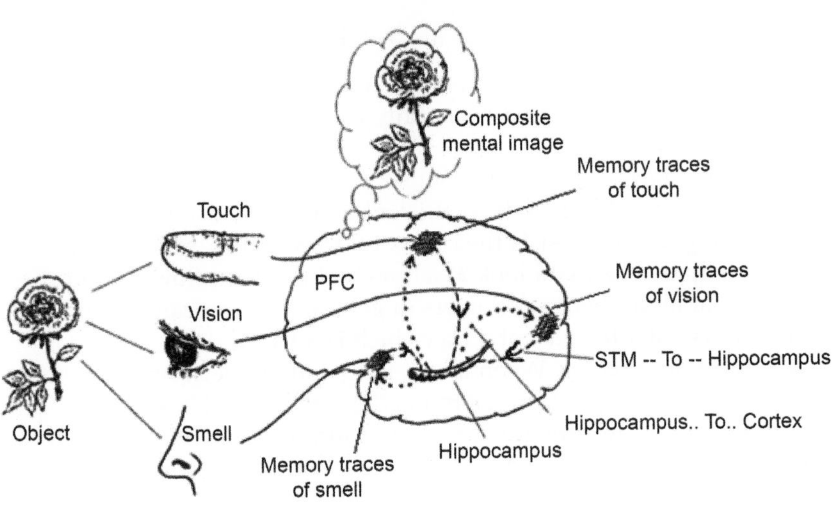

FIGURE 3.6 Mechanism of LTM formation.

in LTM. In the above example, the perceptions of the rose flower (visual, olfactory and touch) that are generated in the respective cortices are sent to the hippocampus, where they are processed and are sent back again to the respective cortices for storage in LTM (Fig. 3.6). This process of storing memory from STM into LTM is called ***memory consolidation***. It has been shown that the hippocampus itself will not store any memory – it will only process the information to be stored as LTM. It is the cortex which is shown to store LTM (Guyton & Hall 2008, p. 557).

It can be presumed that the neurons with their numerous synapses and neurotransmitters take part in memory consolidation. The biophysical and biochemical changes that occur in the neuronal circuits during the process of LTM storage result in the formation of ***memory traces*** (Fig. 3.6). What is the exact mechanism by which memory is encoded (or engraved) in the neurons of the cortex? We will study the neuronal basis of memory briefly below, and discuss it in more detail in Chapters 7 and 8. Before going into the molecular mechanism of LTM, we will briefly study the hippocampus.

Hippocampus

The hippocampus (Ch. 1, Fig. 1.3) is developmentally an ancient structure, and some rudimentary form of hippocampus is seen even in the earliest vertebrates, which assist them in navigating efficiently in their surroundings. In humans and higher animals it is highly developed and assists in the formation of solid LTM.

The functions of hippocampus are as follows:

- ***It is a sorting center.*** It sorts out what is needed for the present situation, letting the cortex store only relevant events useful for future retrieval.
- ***It is an analytical center.*** It groups a new experience with older experiences and hence aids in efficient memory. It compares old and new thoughts, and assists the cortex to draw analogies into *genres*. For example, when you look at a bunch of different flowers you know that all of them are flowers (a genre), and you can say that one of them is not a flower but a fig (which is a fruit).
- ***It is a training center.*** When a task is repeatedly performed or memorized the hippocampus trains the cerebral cortex so that neural circuits are strengthened to perform the task. After some time the task is performed by the cortex, without the assistance of hippocampus. For example, when a student learns his math tables by rote his hippocampus assists the cerebral cortex to memorize and categorize the tables in his LTM, and as the student masters the math tables he will be able to repeat them effortlessly using only his cortical centers (hippocampus not being used any longer). In conditions where hippocampus is damaged (e.g. in Alzheimer's disease or senility), there is difficulty in forming *new* memories, but the memories already stored in the past are left intact. It is a common knowledge that some older people remember events of their childhood well and enjoy telling them repeatedly to their grandchildren (these are old memories), but these grandparents might forget where they have kept their reading glasses (these are recent memories).
- ***It is a 'gate keeper'.*** Only if the hippocampus allows processing of STM is it stored in LTM in the form of memory traces in the cortex – if not, the information fades from the brain. When the same image is experienced again after some time, these synapses are reactivated, enabling recall of memory.

Hippocampus is the **pen** with which we can write our long-term memories – the **ink** being the sensations – and the **paper** being the cortex.

Cellular Pattern of Hippocampus

The hippocampus has only three layers (in contrast to the regular six-layered pattern of neocortex, see Ch.5, p. 106). Glutamate or aspartate (excitatory neurotransmitters; NTs) appears to be the major excitatory NTs in hippocampus, but deeper neurons use GABA (inhibitory NT). There are many neuropeptide-secreting neurons here, such as dynorphin, encephalin, somatostatin and others, which influence bodily functions through the limbic system. In 1949 Paul MacLean suggested that hippocampus forms an *'emotional keyboard'* for the brain.

Trilaminar Loop of Hippocampus

Hippocampus is divided into three areas, CA1, CA2 and CA3, which form a circuit called the *trilaminar loop of hippocampus* (processing center of LTM). The loop is briefly described here. Signals enter via axons of the *entorhinal cortex,* and terminate in the *dentate gyrus*. This forms the first loop. From there neurons of dentate gyrus send mossy fibers to end up on the dendrites of the pyramidal cells of CA3. This forms the second loop. The axons of the pyramidal cells in CA3 divide into two branches – one branch crosses the corpus callosum to reach the opposite side of the hippocampus, and the other branch forms *Schaffer collateral pathways* that connect to CA1. This forms the third loop. From there they leave the hippocampus to return to the respective sensory cortices.

Place cells

It is a common observation that we are well oriented in place and that even if we are left in a dark alley we grope about and find the way. This space-orientation is enabled by pyramidal neurons in the hippocampus called place cells. The surrounding external world is represented in the hippocampus as a *cognitive map* that instructs the cerebral cortex how to move around. This cognitive map is important in the sense that we can search and hunt using it. This enables us to move around in our offices effortlessly, without bouncing into the furniture.

The Papez Circuit

In 1937 James Papez described a neural circuit (Fig. 3.7) for the cortical control of emotions. Since then the circuit has been extensively studied and is now believed to be the major pathway for storing memory. It shows that neural pathways connect the hippocampus with the limbic system, indicating that memories that are stored from the hippocampus have a heavy emotional content. This may partly explain why we tend to remember happy or sad moments of life more readily than 'routine' everyday events. This circuit also explains why some memories also tend to cause autonomic effects; for example, the very memory of the sour taste of lemon may cause us to salivate.

Briefly, the circuit starts with the hippocampus connecting to mammillary bodies (part of the thalamus), through the subiculum and fornix. The mammillary bodies send signals to anterior thalamic nuclei, from there to the cingulate gyrus (Fig. 3.7). The cingulate gyrus sends back signals to the limbic system and to the cortex (to show overt body effects and feelings).

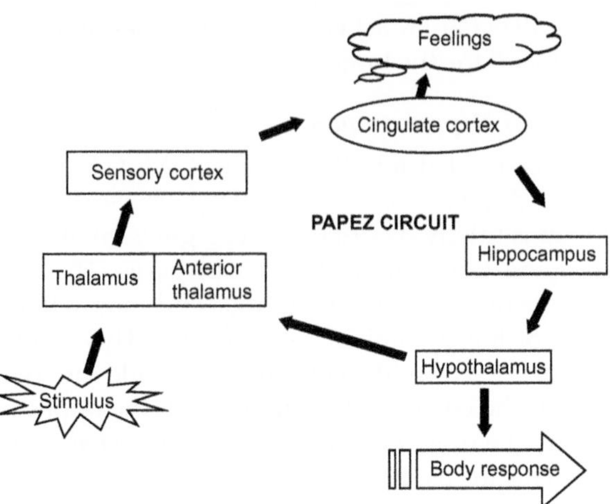

FIGURE 3.7 The Papez circuit.

Lesions of the Papez circuit occur in alcoholics with vitamin B_1 deficiency where the mammillary bodies are affected, leading to an amnesic (forgetfulness) syndrome called *Korsakoff's psychosis* (see above), which is characterized by confabulation, confusion, disorientation and a marked inability to store new information (anterograde amnesia).

NEURONAL BASIS OF MEMORY

So far we have seen how memory is formed at a gross level in the brain – we have seen some of the structures and circuits in the brain that are involved in memory formation. Now we will see the mechanism of memory at the neuron level. The cornerstone of memory is the ***synapse*** – this is because synapses connect one neuron with another, resulting in a ***network of neurons***. We will now briefly study these neuronal networks which are responsible for memory.

Synapses and Neuronal Networks

The sensations we receive travel up to the sensory neurons where they are converted into percepts (or thoughts). Several percepts join together to give rise to a composite image of the external world (Fig. 3.6). However, it must be noted that ***no single neuron*** is capable of holding all the information needed to build up a composite image of the external world (Bruno, 2002). For example, an image of a flower is made up of several colors,

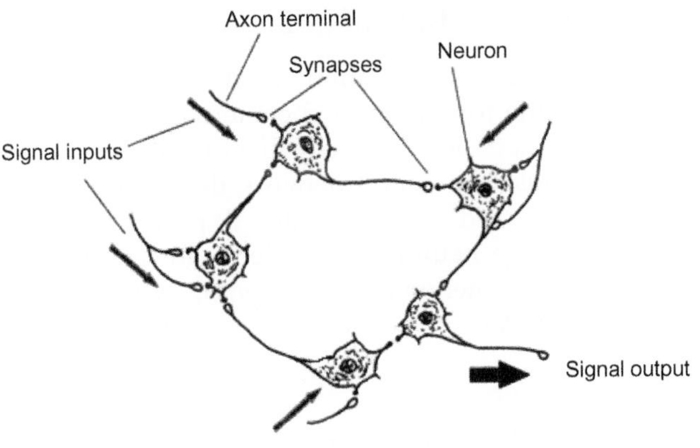

FIGURE 3.8 Neuronal network and signal transmission.

shades, and contours – each of these sensations must be carried in bits to many concerned neurons to build up a meaningful image of a flower – a single neuron obviously cannot hold all the information.

Thus, for this composite image to form, several neurons receiving different types of sensations have to *'cross-talk'* with each other to produce a composite image. The way they talk to each other is via the synaptic connections between them (Fig. 3.8). Thus, to build up a comprehensive image of the external world we need to have synapses between neurons. When the synaptic communication is strengthened these neuronal networks become strong; and they become weak when the synapses are weakened (Bruno 2002). Consequently, the durability of the memory traces depends on the strength of synapses involved in the neuronal networks.

Synaptic plasticity (see below) is the ability of neurons to bring about changes in the connections between neuronal networks in response to use or disuse. It has been shown that structural changes occurring in synapses are responsible for the development of LTM (Guyton & Hall 2008, p. 725), and at present, synaptic plasticity is the hot topic of all the researchers concerned with memory. There are essentially two ways by which synaptic plasticity can take place:

- By the NTs (which is being extensively studied; see p. 68)
- By dendritic spine plasticity (which is not very well known; see p. 70).

It is not clear how exactly these changes would bring about changes in memory, but it appears that these mechanisms act in conjunction in the process of memory formation. We will study these mechanisms briefly here. However, they are further discussed in Chapters 6–8 as a part of *'molecular-grid model'*.

Synaptic Plasticity: The Role of NTs

When information reaches the brain, a network of neurons becomes excited, resulting in the formation of short-term memory. Once STM is formed by a neuronal network, it is likely that the memory will fade away after some time because the neural network that is responsible for this memory may be 'disbanded' sooner or later. To prevent this, the synaptic connections between the neurons concerned with memory have to be strengthened. This phenomenon of strengthening synaptic connections is called *long-term potentiation*.

Long-Term Potentiation

NTs are chiefly involved in the development of long-term potentiation (LTP), though the role of dendritic spine plasticity (see below) is emerging as a prominent player owing to current research in memory and learning. ***Glutamate*** is the chief excitatory NT responsible for long-term storage of memories – 75% of excitatory transmission in CNS is mediated by glutamate. Glutamate acts by binding to the *glutamate receptors* present in the postsynaptic membrane, causing depolarization of the dendritic membrane (see Ch. 2, p. 49).

LTP occurs chiefly in the hippocampus and other regions of the cerebral cortex. Cerebellum and basal ganglia also show evidence of LTP; perhaps this is to mediate motor memory.

Glutamate Receptors

Glutamate (released by axons) binds with glutamate receptors (on the dendritic membrane) and excites them. These receptors, in turn, activate the ion channels in the dendritic membrane, and this causes an ionic shift across the membrane, resulting in depolarization (see Fig. 2.9).

The glutamate receptors are of two types. We will briefly learn about them without going into the details.

AMPA Receptors

When glutamate binds to these receptors the ion channels are activated, which results in influx of Na^+ ions causing reversal of polarity (*depolarization*) of the dendritic membrane (see Fig. 6.4). This ionic flux generates membrane potentials of different amplitudes (*graded potentials*, Ch, 2, p. 40), which propagate along the dendrite to the soma (body), and from there to the axon hillock. Several such graded potentials are needed to excite the axon hillock (see Fig. 2.5). Finally, when the axon hillock is excited, then an *action potential* (AP) is generated. This AP propagates along the axon to reach for another neuron's dendrite. This is the way information propagates from one neuron to the other, until the entire neuronal network gets excited.

NMDA Receptors

This is a bit tricky. This receptor is associated with a calcium channel, but is constantly blocked by a *magnesium ion* (Mg^{2+}), so that even when glutamate binds to these receptors it fails to open the associated calcium channel. For the NMDA receptor to really open its calcium channel the dendritic membrane has first to be depolarized. This is done by AMPA activation (as described above). When NMDA receptors are finally activated, they allow a huge influx of Ca^{2+} ions into the dendrites, which causes rapid depolarization. This means that for the NMDA receptors to work they have to be first primed by the AMPA receptors – then their total effect can be really big. The influx of Ca^{2+} ions induces several intracellular events for prolonged periods, thus leading to LTP. If NMDA receptors are blocked by drugs then LTP cannot happen, so that LTM cannot be achieved.

Current research shows that, apart from the above glutamate system of excitation, there are several other major *neuromodulatory systems* involving molecules such as *dopamine* and *serotonin*. These molecules mediate LTP and help in various complex activities such as motivation, reward, instinct, emotional behavior, etc., but these mechanisms are unclear.

Second Messenger System

This is an intracellular mechanism which helps in storing LTM for long periods. This works by bringing about certain changes in the *genes* thereby changing the *proteins* involved in memory storage process. The second messenger system is mediated by calcium ions. The proteins synthesized by this mechanism also influence the formation of dendritic spines, hence it also takes part in dendritic plasticity (see 'Synaptic plasticity', below).

Hebbian Process

In 1970, Donald Hebb suggested that if two neurons are active at the same time then the synapses between them are strengthened (the Hebbian process). Hence the old catch-phrase:

> Neurons that fire together, wire together.

It has been shown that when the pyramidal cells of hippocampus are excited by a sustained stimulus they remain excited for long periods (perhaps for several weeks) and form new connections. This is shown to be very important in the formation of LTM. A variant of the Hebbian process is discussed in Chapter 8 ('Formation of dendritic pleats' p. 155).

Long-Term Depression

Long-term depression (LTD) is the opposite of LTP, and is characterized by a decrease in postsynaptic strength. This happens by dephosphorylation of AMPA receptors and the facilitation of their movement away from the synaptic junction. LTD is prominently seen in the hippocampus and cerebellum, though current research shows LTD in other areas of the cortex involved in memory and learning. This mechanism is chiefly responsible for the removal of old memory traces (facilitating the formation of new memory traces). However, LTD also sharpens the image (or any sensory perception) by enhancing the contrast. It also has a major role in executing motor memory (for example, riding a motor bike without conscious attention).

Synaptic Plasticity: The Role of Dendritic Spines

The classic synapse is the axodendritic synapse, i.e. an axon ending on the dendrite, which means that axon's membrane forms the presynaptic membrane, and the dendritic membrane forms the postsynaptic membrane (Fig. 2.7). These axodendritic synapses are far more common in the CNS than other types of synapses (Ch. 6, p. 110), hence in the following discussion it is presumed that axons send signals and dendrites receive them.

Synaptic plasticity mediated by dendritic spines is a relatively ill-understood topic, but at present it occupies the forefront of research in memory and learning. This type of synaptic plasticity (called '***dendritic plasticity***') is again studied in Chapters 6 and 8 as a part of the '*molecular-grid model*'. However, a brief account of dendritic spines is given below.

Dendritic spines are pseudopodia-like structures, with a length of about 0.01–0.8 μm, projecting from the dendrites. In fact, dendritic spines are the places where the axons come in contact to make a synapse. Current research shows that they are highly dynamic structures which make or break connections with axons constantly throughout the learning period (Guyton & Hall 2008, p. 725). It appears that each incoming stimulus induces the dendrites to move towards the axons and to form new dendritic spines over them to make new synapses (see Figs 3.9 & 6.3). This property of dendrites gives the neuron its enormous capacity for plasticity, because synaptic plasticity is the ability of neurons to bring about changes in the synaptic connections.

How are these movements of dendrites and formation of spines possible? The dendrites have abundant protein strands in them, called cytoskeleton (see Ch. 2, p. 42), which are highly dynamic. The rearrangement of these cytoskeletal strands is responsible for the growth and movement of dendritic spines towards the stimulus.

The number of dendritic spines steadily increases from the fetal stage to adult life, which signifies their role in learning and memory. In adult life, there is some evidence that the structure of the dendritic tree itself

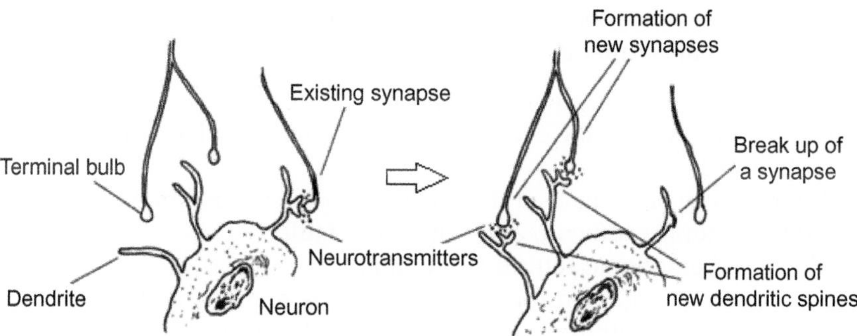

FIGURE 3.9 Synaptic plasticity – the role of dendritic spines.

remains almost unchanged (see *pruning* in Ch. 6, p. 115), but the level of neuronal plasticity remains fairly high even into old age (indicated by the human ability to learn even as age advances) owing to the versatile ability of dendrites to form new dendritic spines.

Current research clearly shows the indubitable versatility of dendrites in human intelligence. It has been shown that in children with learning disabilites, such as Down syndrome and Fragile X syndrome, the dendritic spines show poor development (Wang et al. 2012). It is also shown that the number of synapses has no relationship with plasticity – in fact, it is established that too many strong synapses makes learning difficult, whereas fewer synapses make learning easier, presumably because strong synapses do not permit plasticity, and plasticity is more important for learning than the number of connections.

FACTORS AFFECTING MEMORY

The human memory is reasonably accurate: consider a student memorizing his lessons; a doctor remembering signs and symptoms of rare diseases; an engineer applying his theorems at his work; all of which speak for the perfection a human being can achieve in memorizing details. However, there has been some debate on how perfect human memory can be; for example, forensic psychologists are concerned with the accuracy of the memory of bystanders in the description of details of a crime – but that is not a precise test of the bystanders' memorizing capacity, because memory is dependent on many factors. Einstein was known to be forgetful of dates, phone numbers and birthdays, but this does not imply he had a poor memory, it only means that he was possibly less attentive to those less deserving aspects of life, and perhaps this gave him greater brain space for more deserving scientific concepts. The following are the factors which affect memory:

1. Alertness, attentiveness and vigilance.
2. Interest involved in the event, which is again dependent on need and motivation.
3. The presence or absence of distractions; a preoccupied mind makes memory storage difficult, and sometimes faulty.
4. Mood and emotional setting at the time.
5. Surroundings, e.g. disturbing noise makes remembering difficult.
6. Memorizing something in relation to a familiar fact is easier. Students can remember better if they can associate new memories with old memories. Live demos and visual aids are helpful to students in this way.
7. Memorizing in relation to a given situation (*contextual memory*) is very helpful.
8. Children's nutrition is also important, because an enormous amount of protein is necessary to build new memories (see Ch. 7, p. 140). The importance of fats in the diet (e.g. DHA) is also well known (see Ch. 8, p. 161).
9. A good night's sleep gives you a good memory; there is evidence that sleep facilitates the consolidation of memory in the brain.

SOME PHENOMENA RELATED TO THE BRAIN

The human mental faculty is truly amazing. We will describe here a few of the phenomena that are associated with the brain, though we have not yet fully understood their physiological mechanisms. It should be noted that memory is the cornerstone of all these phenomena.

Language

Language has a fundamental role in human intelligence and is the key part of human culture. Two important centers are identified for the production of language (Fig. 3.10). One is the sensory area called *Wernicke's area* (in the temporal lobe of the dominant hemisphere, usually the left), and the other is the motor area called *Broca's area* (inferior end of motor cortex) (see Ch. 1, p. 11). Wernicke's area is involved in understanding written and spoken language and Broca's area in speech production.

For speech to be produced the sensory signals have to be first processed in Wernicke's area. For example, a pin prick to the heel may reflexly cause you to say 'ouch!'. For this vocalization to happen the sensation of pain must pass first to Wernicke's area, where the sensation is recorded, and

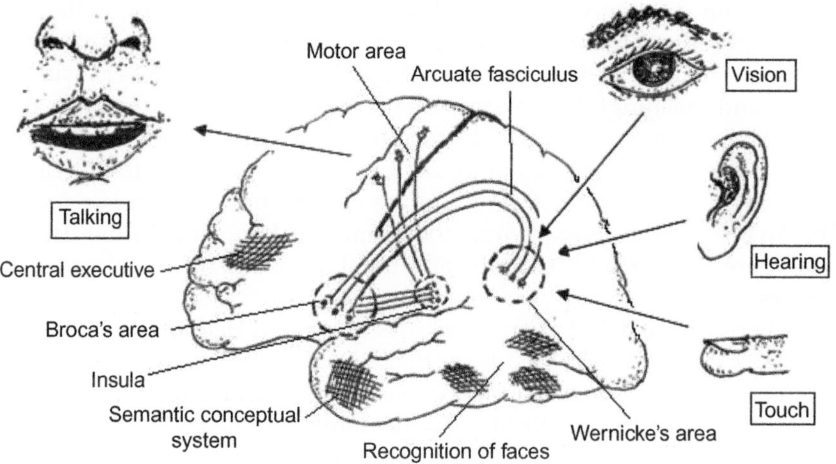

FIGURE 3.10 The language pathway.

from here the signals are routed to Broca's area through a bundle of fibers called the *arcuate fasciculus*. Broca's area processes these signals and relays them to the motor cortex, which instructs the various muscles of the lips, tongue, larynx and other structures to produce *articulation* and *vocalization* (Fig. 3.10). Similarly, if you look at a rose and want to say 'rose', the visual signals have to be processed first in Wernicke's area, which collects the old memory of rose and forwards it to Broca's area. A similar transformation has to take place with all sensory modalities.

It is interesting to note that when adults learn a second language, a second Broca's area is created adjacent to the original one concerned with the mother tongue; whereas when a child learns a second language along with their mother tongue, there is a single Broca's area (Barrett et al. 2012, p. 293). Hence, children acquire fluency in a second language more easily than adults. If there is defect in these areas it produces a speech defect called ***aphasia***.

Sleep and Dreams

Sleep is an essential prerequisite for health, and humans spend about one-third of their lives sleeping. Sleep occurs in two forms – ***rapid eye movement*** (REM) ***sleep*** and ***slow*** (non-REM) ***sleep***. REM sleep generates a rapid, low-voltage electrical activity in the brain, and the characteristic feature of this sleep is the occurrence of rapid, roving eyeball movements. The waves in REM sleep resemble that of an alert state. Dreams happen in REM sleep, which takes up 25% of a human's sleeping hours. Non-REM sleep records slow but high-amplitude electrical waves, and is not associated with eye movements.

The main difference between REM sleep and wakefulness is that of meaningfulness of thoughts – dreams during REM sleep are characterized by bizarre imagery and illogical thoughts, whereas wakefulness is usually logical and meaningful. The dream state shows increased activity in the pons, limbic areas and visual association areas, which together form a *'closed neural system'* (Barrett et al. 2012, p. 275). This means that dreams are drawn from past memories, and are cut off from the sensory areas. The limbic system gives dreams their emotional content. In the dream state there is *decreased* activity in prefrontal, parietal and visual primary cortices.

Perhaps during REM sleep, memories are reactivated and put into the right perspective. There is evidence that a good night's sleep consolidates memory and improves performance significantly – thus a fitful sleep gives you a good LTM. During REM sleep the noradrenaline levels in the brain (associated with stress) show a sharp decline.

Déjà vu and Other Phenomena

Occasionally, as you walk along a strange dark corridor, for example, you may experience a feeling of peculiar familiarity with the surroundings-even though you know that you have not visited that area before, you may feel that you *have* seen that place in your past. This false sense of familiarity is called *déjà vu phenomenon* (in French, 'already seen'). Stimulation of part of the temporal lobe areas of the brain is often associated with this sort of alteration of the interpretation of the individual's surroundings – and there may be a role for *place cells* (see above) in this phenomenon. Déjà vu may occur sometimes as an aura (premonition) in temporal lobe epilepsy. Sometimes an opposite feeling of strangeness is felt in familiar surroundings.

There are many other perplexing cognitive phenomena, such as illusions, hallucinations, hypnosis, synesthesia, strange mental disorders etc., which need to be discussed in relation to the phenomenon of memory, but these vagaries of the human mind are beyond the scope of this book.

Having studied the anatomical and physiological aspects of the human brain in the preceding chapters, we are now ready to examine the perplexing problem of how thoughts are generated by the neurons in Chapters 4–8.

PART II

THE MOLECULAR-GRID MODEL

C H A P T E R

Primary Thoughts and Ideas

OVERVIEW

In Chapters 1–3 we studied the anatomical aspects of the brain, the functional aspects of neurons, and considered the human memory. However, nothing was made clear regarding the ***basic properties of thought*** in those chapters – thought itself has remained an abstract phenomenon.

Chapters 4–8 undertake a scientific examination of the study of thought, which finally shows us the exact mechanism of generation of thought by the neurons. The approach is, as the reader can see, entirely logical, deductive and methodical. Before starting with the study, let us first put forward the following three questions:

1. What is the basic unit of thought?
2. Where is it generated?
3. How is it generated?

In this chapter, we will examine ***what*** constitutes the basic unit of thought and what are its fundamental properties. Chapters 5 and 6 are concerned with the study of ***where*** exactly in the neurons the external stimuli are converted to give rise to thoughts. Chapter 7 and 8 are concerned with the study of ***how*** exactly are these thoughts generated by the neurons.

CONCEPT OF PRIMARY THOUGHT

The first step in the generation of thought is that the reception of a ***stimulus*** from the outside world. A stimulus excites the receptor and the resultant message is transmitted as a nerve impulse to the neuron (see Ch. 2, p. 31 & Fig. 2.1). The receptor sends the signal as a ***sensation****, and

* Throughout the book the sensory signals that arise in the peripheral receptors, such as eyes, ears, skin, etc., are called sensations for the lack of a better term. This is to differentiate sensations from perceptions, which are the signals that are perceived in the brain

The Biology of Thought.
DOI: http://dx.doi.org/10.1016/B978-0-12-800900-0.00004-X

the neuron converts it into a ***perception*** or ***thought*** (see Ch. 3, p. 54 & Fig. 3.1). For example, the cones in the retina (receptors for color sensation) receive a 'sensation of color'; and the sensory neurons of the visual cortex convert this into a 'thought of color'. For a receptor it is only a sensation, whereas for a neuron it is a thought. This initial unit of perception or thought at the level of neuron is called the ***primary thought***. These primary thoughts are the fundamental units of the human mind, and form the building blocks of more complex ideas. The brain works with several of these primary thoughts to build up an idea of the external world, as will be seen.

This fundamental unit of thought in the brain also forms the basis of memory.

The basic properties of a primary thought are examined below under the following headings:

- Primary thought as a fundamental unit
- Primary thoughts and memory
- The sensory experience is the same irrespective of stimulus
- We can only sense what our mind knows
- The non-innovative nature of human thought
- Inheritance of primary thoughts.

PRIMARY THOUGHT AS A FUNDAMENTAL UNIT

Primary thoughts are the most fundamental units of perception which are unitary, non-comparative, non-associative and *a priori* experiences.

What Is a Primary Thought?

Let us first define a primary thought. In order to learn what exactly is a primary thought we will conduct a thought-experiment. Place a ***perfectly uniform blue colored sheet*** just in front of your eyes, so that nothing else except the blue color is seen – the eyes see only a monotonous blue color in all directions and nothing else. Now, what do you perceive? What is the ***first perception*** that comes to your mind? Your mind first receives only a sensation that has no feature and no name, because there is only a nondescript uniform color in front of you which has no feature, and you have not yet named the color. Your mind receives this 'unnamed color' ***not*** from your experience, but from the basic structure of the mind itself. This is called the ***primary thought***.

Such primary thoughts are formed from each and every sensory input – from both the general and special senses, such as sound, smell, taste, touch, pressure, temperature, etc. (see below), but only for simplicity's sake is a sensation of color taken in the above thought-experiment. The neurons convert all types of sensory inputs from receptors into primary thoughts – light signals into perceptions of light in the visual cortex; auditory signals into perceptions of sound by the neurons in the auditory cortex; odor signals into perceptions of smell by the olfactory neurons; tactile signals into perceptions of touch by the parietal cortical neurons, and so on. Thus, a primary thought is the basic unit upon which our ideas of the external world are based, as shown below. The primary thoughts thus formed are stored in the brain, and this is the first step in the formation of memory.

What is an Idea?

Now, in the above thought-experiment, we can name this unnamed color as 'blue' because we were taught to assign a name for this color in our childhood. The notion of color with a name is an assembly of primary thoughts, because such an idea of blue color is produced by combining the primary thought of the unnamed color generated in the visual cortex with the primary thoughts of sound signals generated in the auditory cortex (for full explanation see below). This assembled notion is called an ***idea***.

Now we have formed an idea that the color is blue. Continuing with our thought-experiment, we may say that the endless blue color which we have perceived may be one of the following things, basing on our past experience:

- We may say that it is the vast blue sky of a clear cloudless afternoon
- Or, perhaps it is a vastly expansive blue wallboard just in front of our eyes
- Or, that it is an immense expanse of sea
- Or, something else of that kind.

The idea of a 'blue sky' or a 'blue wallboard' or 'blue sea' depends on the previous experience and preference of an individual. These ideas are more complex than the idea of color blue (as shown above), because they involve many more primary thoughts and ideas. These complex ideas are derived from a combination of many primary thoughts and ideas drawn from our long-term memory and learning. Our day-to-day experiences involve all very complex ideas – a seemingly simple object such as a football we look at is the product of several primary thoughts; likewise the flavor of our favorite dish, the sound of music and many such others are all assembly of primary thoughts (i.e. complex ideas). The mechanism of formation of simple and complex ideas in the brain is described in Chapter 5, pp. 102–105. For now, we will simply study the properties of primary thoughts and ideas.

Characteristics of Primary Thoughts

Primary thoughts differ from ideas in many respects. We will first discuss the characteristic features of primary thoughts, and then those of ideas.

Primary Thought is Unitary

Primary thought is carried as *single unit* and forms the fundamental unit of mind. This constitutes the *smallest* or the *most basic* unit of thought. In the above example, the unnamed, featureless color is the primary thought; for other examples see Examples of primary thoughts and ideas, below.

Primary Thought is Indescribable

Any attempt to describe a primary thought leads to the formation of an idea, hence primary thoughts are exquisitely sensed by the mind and cannot ***be expressed or described in words*** in the truest sense; for example, we cannot exactly describe in words *'what is blue'* – the sensation has to be exquisitely experienced by the individual. Primary thoughts are known *subjectively* – just by the individual himself. In the same manner, all the primary thoughts of smell, taste, sound, pain etc. are subjective and are thus indescribable; for example, how can we describe what sweetness is? How can we describe what sound is? Sensation of pain is also subjective. How can anybody describe the magnitude of pain that only they can feel?

Note: However, indescribability *itself* will not entitle the sensation to be a primary thought, because such indescribability is a feature of many experiences we sense and feel, such as happiness, sadness, pleasure, anger etc. How can anybody describe what is happiness, for example? But we know that these experiences are not primary thoughts, but highly complex ideas.

Primary Thoughts are Non-comparative

As seen above, our sensation of blue color can be exclusively known to us. It may appear that our fellow human beings also perceive an exactly similar sensation for blue color, but there is no means by which we can compare our sensation with others and hence we *cannot* be sure of the fact that all the others see the color blue in the same way as we do (see 'Note' below). We will shortly see that ideas are comparative (see below). This property is also true of all other primary thoughts related to both general and special senses – all the basic units (i.e. primary thoughts) of our taste, our smell, our touch or our pain cannot be compared. Hence primary thoughts are non-comparative (Table 4.1).

Note: In all probability, it can be presumed that we all perceive sensations *slightly* differently from others – what we see as blue color is likely to appear slightly different to our neighbor, what tastes like sugar to us may be slightly different for the next individual, so on. This presumption

TABLE 4.1 Primary Thoughts

PRIMARY THOUGHTS ARE:
Unitary
Indescribable
Non-comparative
A priori experiences

can be substantiated by the fact that we all have subtle differences in our physical features, such as fingerprints, *iris recognition system, facial patterns, voice patterns* etc., so why should our senses match each other exactly? Thus, all our thoughts must be highly individualistic. The reader may get a better perspective on this matter later in this chapter (p. 90) and in Chapter 7, p. 140, which offer a genetic explanation for this feature.

Primary Thoughts are* a priori *Phenomena (Existing Prior to Experience)

Primary thoughts are built in the fundamental structure of the mind, and thus they need no prior experience. To sense an external world event (for example, to sense a blue ray of light) the external sensation has to be converted into a primary thought by the neurons – and for this conversion to take place some sort of *molecular mechanism* must be present somewhere in the neurons (discussed below). These molecular mechanisms or ***'molecular gadgets'*** (as we may call them) must already be lodged in these neurons in order to convert sensations into primary thoughts, and if they are absent the sensation cannot be converted into perception. As an example, an infant can perceive the unnamed color (though it has not yet learned to name the color) because it houses the necessary molecular gadgets in its neurons to decipher this sensation. We may also consider this – human beings *cannot* perceive ultraviolet rays because we lack the necessary molecular gadgets in the nervous system (see also 'Primary thoughts and molecular gadgets' in p. 94, below). Thus, the perception of primary thoughts needs no past experience and this capability of perceiving a primary thought is innate. This is an *a priori* experience (existing prior to experience). A full explanation of this property is given in 'Primary thoughts are inherited', (p. 90), and in Chapter 7, p. 140.

Characteristics of Ideas

The following are the characteristics of ideas. The reader may appreciate the differences between primary thoughts and ideas by making a comparison as he/she reads along.

Ideas are Modifiable

An idea is an *assembly* of primary thoughts. As demonstrated above, a group of primary thoughts join together to form an idea. In the above thought-experiment, the unnamed color is a primary thought, and when we have associated it with the name 'blue' it becomes an *idea* (see above). We will now see how an idea can be modified. The *idea* that the color is *'blue'* changes from language to language; for example, a person who knows only Hindi associates this unnamed color sensation by the auditory signal *'neel'*, whereas a person whose only language is German knows this by *'blau'*. Thus, ideas are ***modifiable*** – we may join this unnamed color with some other auditory signal to give rise to another idea. The primary thought itself has no language, but when we attach a meaning to a primary thought it becomes an idea.

In the above thought-experiment the same uniform blue color may be perceived as *'a vast blue sky'* by one individual, or *'an expansive blue wallboard'*, or *'vast sea'* by other individuals. It means that the same set of primary thoughts may give rise to different ideas, depending on our past experience or memory. If we break up an idea into small units we will find that an idea is only an *ensemble* (a group) of several primary thoughts. We can associate one primary thought with other primary thoughts to form a countless number of new ideas. Thus, the primary thoughts may be limited in number, but the number of ideas is unlimited. The primary thoughts may be comparable to ***pixels*** in a digital camera, i.e., the number of pixels (= picture elements) generated by a digital camera may be of a limited variety, but the camera can reassemble these pixels to produce an unlimited number of digital pictures (see explanation in 'We can only sense what our mind knows', below).

Ideas are Describable, Comparative and Associative

We can describe an idea in words to our fellow human beings (compare this with non-comparative primary thoughts, as discussed above); (Table 4.2). As an example, we can look at a multicolored picture of a bird painted on the wall (which is an example of a complex idea) and we can

TABLE 4.2 Ideas

IDEAS ARE:
Modifiable
Describable
Comparative
A posteriori experiences

compare our mental picture of that painting with that of our friend – we can also paint an exact copy of it on a paper for demonstration. In fact, earning our daily bread depends much on our ability to demonstrate our ideas to others. If we 'break up' these ideas into small perceptions, we may find that they are composed of several primary thoughts and each one of them is again non-comparative (as described above).

Ideas are a posteriori Phenomena:

As learning progresses, new ideas are woven from primary thoughts and stored in the memory for easy recollection (Ch. 8, p. 153). We can build innumerable number of new ideas by using various combinations of primary thoughts. These are all *a posteriori* phenomena (derived from experience). See the discussion on how innovative is human thought, below (p. 88), for a full appreciation of this phenomenon.

Examples of Primary Thoughts and Ideas

It is very difficult to always isolate primary thoughts from ideas for all the sensory inputs we experience – it is not possible to differentiate one from the other at all times. However, we will now describe a few primary thoughts which stand out as clear-cut examples of primary thoughts:

- The primary colors we sense, ***red – green – blue***, are perhaps the best examples of primary thoughts. The sensation of white is the result of a conglomeration of all colors, hence ***white color*** is an ***idea*** (Note: there is nothing called 'white' in nature, it is only a human perception – humans perceive VIBGYOR [violet, indigo, blue, green, yellow, orange, red] colors *together* to give the sensation of whiteness, as demonstrated long ago by *Newton's color wheel* experiment). The secondary colors such as brown, indigo, violet etc. are also ideas.
- The ***special senses***, such as the basic units of hearing, smell and taste, are all certainly unitary primary thoughts. For example, the ***basic tastes*** we appreciate (sweet, bitter, sour, salty, and umami) are all primary thoughts, and are *a priori* experiences. The various ***flavors*** we relish at our dining table are merely complex ideas that occur because of the amalgamation of several of these primary thoughts of taste along with primary thoughts of smells – these are all really *a posteriori* ideas.
- The ***general senses***, such as touch, pain, temperature, etc., are perhaps other good examples of primary thoughts. We can see that all these sensations can be perceived, but cannot be expressed in words and cannot be compared. They are indescribable because they are subjective phenomena.

- The perception of a ***note*** in a piece of music is perhaps a primary thought. Several such notes assemble to give a ***musical notation*** (which is an idea). The whole of the '***raga***' is a complex idea.

PRIMARY THOUGHTS AND MEMORY

In this section it is shown that primary thoughts and human memory are interdependent, and that primary thoughts cannot be generated without the existence of innate memory.

Primary Thoughts and Molecular Gadgets

This is an important feature of primary thought. Take, for example, the uniform blue color in the above thought-experiment. The cones (receptors) send impulses to neurons in the brain where they are deciphered first as an unnamed color. The virgin mind of an infant can also perceive the unnamed color. How is this possible? This is because the basic structure of the human brain has the necessary ***molecular gadgets*** (see above) in their neurons to decipher the received sensation of this unnamed color from the external world. A molecular mechanism was already stored somewhere in the neurons to sense this unnamed blue color. In other words, the neurons have an ***innate memory*** of this color, and without this prior memory the neurons cannot perceive this color blue (see examples below to understand this property better). In the absence of this ***a priori memory*** (=existing prior to experience, see 'Primary thoughts are *a priori* phenomena', above), the unnamed color sensation transmitted by the receptor fails to excite the neuron, and thus we become 'blind' to that particular sensation.

A sender becomes useless in the absence of a receiver.

Compare this with the following analogy: a light bulb glows when electricity is supplied; the filament in the bulb makes all the difference. In the absence of a filament the bulb fails to glow even when electric current passes through it. Likewise, a sensation from a receptor becomes null-and-void in the absence of a perceiving molecular gadget in the neuron.

Note: What are these molecular gadgets, and where are they located? This is discussed in the subsequent chapters, but for now we will study the characteristics of primary thoughts and ideas exclusively.

Primary Thought and Memory are Interdependent

Now it is understood that there can be no primary thought without *a priori* memory – or simply put – there is ***no thought without memory***, because for a primary thought to form, a pre-existing molecular gadget is mandatory, as seen above. Hence, it can be concluded that thought and memory are interdependent – thought cannot operate without a stored innate memory. A full explanation of this phenomenon of innate memory is presented in Chapters 7 and 8.

A Few Examples

- Take again, for example, ultraviolet rays in human beings. Humans cannot perceive UV light (or X-rays, or gamma rays, for that matter) because they do not have the necessary molecular gadgets to decipher them; hence these rays are invisible to them. Whereas some animals, such as birds, bats and bees, have the molecular mechanisms to perceive UV rays, and they utilize UV light for efficient navigation.
- In the case of hearing ***sounds***, the human ear is designed to hear sound signals ranging from ***20*** to ***20000 Hertz***, and any sound outside this frequency is not heard by the human ear. This is because the human nervous system is not equipped with suitable molecular gadgets that can receive and convert them into primary thoughts of sounds outside this range. On the other hand, an elephant can hear ***infrasonic sounds*** and bats can perceive ***ultrasonic waves*** – because they are equipped with molecular mechanisms that can decipher these wave lengths.
- The same thing is true with the ***myriad smells*** a dog can sense – a dog has more than 200 million receptors in its nose, in contrast to 10–20 million receptors in humans (the olfactory lobe in dogs is 40 times larger than in humans). Evidently what a dog can smell easily can make no sense at all to a man, because humans do not possess the necessary molecular gadgets to encode all these odors – and imagine, our lives would be really miserable if we were able to smell all the odors that way!

SENSORY EXPERIENCE IS THE SAME IRRESPECTIVE OF STIMULUS

In this section it is shown that neurons generate only their assigned primary thoughts whatever the nature of stimulus.

Explanation

A significant feature of sensory neurons is that the neurons generate only those primary thoughts for which they are designed; for example, neurons concerned with vision can generate only light perceptions; olfactory neurons generate only smell perceptions; auditory neurons generate only sound perceptions – so on and so forth. In other words, the stimulus from the receptors may be in any form, but the excited neurons generate only those primary thoughts in which they specialize. This needs some explanation. The following example of ***phosphenes*** serves our purpose well.

One Interesting Phenomenon is That of Phosphenes

The receptors in our sense organs, such as our eyes, ears, skin, etc., are generally stimulated by their assigned stimuli – eyes are usually stimulated by light waves, ears are stimulated by sound waves, or skin nerve endings are stimulated by touch, pressure, heat, etc.

But we may consider the following situation: our sense of vision arising from the retina can also be stimulated by a stimulus other than light. For instance, when we rub our closed eyes we see sparkles of white light – this phenomenon is called ***phosphenes*** (Fine 2008, p. 50). This happens because the rubbing action builds up pressure inside the eyeball, which falls upon the retina and so the cones and rods in the retina are stimulated by ***mechanical pressure,*** even though *light energy* is *absent*. This pressure energy generates electrical potentials in the rods and cones and these impulses are transmitted to the neurons of the visual cortex, but these neurons perceive the incoming potentials ***not as the pressure sensation,*** but rather they ***perceive the sensation as light perception*** – this makes us see stars of light'. The reader may compare this phenomenon with the '*Labeled line principle*' described in Chapter 6.

The neurons of visual cortex are *not* designed to sense pressure, because they possess the molecular gadgets to generate primary thoughts of light only – thus they can read all incoming stimuli as sensation of light only. Consequently, when pressure is applied to the retina several red, green and blue-sensitive cones are simultaneously stimulated and these sensations reach the brain, and *collectively* they produce the perception of sparkles of white light.

Why are There so Many Receptors?

Now, here comes a question. When we say that the neurons themselves perceive all the sensations, and receptors simply transmit signals, then why are there so many varieties of peripheral receptors? We have *Pacinian corpuscles* for receiving pressure sensation, *Meissner's corpuscles* for sensing touch, *free nerve endings* for sensing pain, *hair cells* to receive sound, rods/cones to receive light, and so on. In the above example of phosphenes, the neurons of the visual cortex perceived the sensation of light, even though pressure had excited the rods and cones – then what is the function of rods and cones? If visual neurons themselves discriminated the light perceptions then we could have only simple free nerve endings, without an elaborate mechanism of rods/cones. Likewise, a single type of receptor would have been sufficient for all other sensations. But, as we can see, there are several designs of receptors. Why?

The reason for the presence of different receptors is that these receptors have a ***minimum threshold*** for their stimulation for a given stimulus. As an example, the rods are excited by the presence of a minimum amount of light, but they need a lot of mechanical pressure to excite them. This property is called '***differential sensitivity of receptors***' (Guyton 2008, p. 572). This means that the rods are designed to have a minimum threshold for light. Similarly, when we touch the skin, the Meissner's corpuscles are excited preferentially, sending sensations of touch, but not the Pacinian corpuscles (which send sensations of pressure) – because Meissner's corpuscles are designed to get excited at the least amount of *touch* stimulus applied to skin whereas the Pacinian corpuscles are designed to get excited at the least amount of *pressure* stimulus. This facility naturally improves our range of perception of the external world; these receptors allow us to discriminate between slight touch and pressure, between pressure and temperature, between light and sound, and so on. In the same way, there are three types of cones to receive the colors red, green and blue – each of these types has a different threshold for different wavelengths of light, to enable us to efficiently discriminate between various colors, giving us the view of a colorful world around us.

We can consider a unit of receptor and neuron as a ***modem*** (much like that of a computer). The receptors may be considered as '***modulators***' because they receive signals from the external world and convert them into electrical signals and pass them up to the neurons. The neurons may be considered as '***demodulators***' because they convert these signals back into perceptions. Now it can be said that the receptors and neurons act like a *modulating-demodulating unit*, and this unit may be called a '***modem***'.

WE CAN ONLY SENSE WHAT OUR MIND KNOWS

The human nervous system cannot understand a sensation for which it is **not** designed as shown here in this section.

Explanation

It has been shown, above, that for a sensation to be experienced there must be a neural mechanism (in the form of molecular gadgets) to sense the experience. Consequently, it can be said that we cannot sense any more primary thoughts than those with which we are endowed (see (pp. 90–92), below). For example, a human being cannot perceive ultraviolet rays, unlike some creatures such as birds and bees. We cannot hear ultrasonic waves as bats do; neither can we hear the infrasonic waves of an earthquake, as elephants do. Consider this – how can we *learn* to sense any more smells than we can already smell? Thus, the human nervous system *must* have a specific molecular mechanism to sense that particular sensation *before* it can actually experience the sensation. Another example: a person with *blue-yellow color blindness* cannot perceive blue color at all, and when blue color is shown to him or her, it is perceived as green. This color blind person is totally blind to the color blue, and completely lacks the knowledge of what blue looks like.

The conclusion is that the primary thoughts themselves are limited in number, innate in nature, and cannot be added – we can only assemble and reassemble primary thoughts into different patterns of ideas (like the limited number of color pixels in a digital camera producing an unlimited number of pictures; see 'Ideas are modifiable', above).

Note: However, it is a common observation that we can add a vast number of new *ideas* with varying complexities throughout our lives, from infancy to ripe old age – we can add new names, new faces or various other new things to our memory; we can enjoy a new piece of music; we can taste new flavors, etc. Actually what we do is that we spin new ideas with the help of the already available primary thoughts, thus the number of new ideas we are able to create is virtually unlimited. But the number of primary thoughts themselves that we can generate is limited (see Ch. 7, 'Genetic basis of primary thoughts', for a full explanation).

HOW INNOVATIVE IS HUMAN THOUGHT?

The things we experience from the external world cannot be totally new, but are simply the product of things experienced by us in the past as demonstrated here.

Explanation

Our primary thoughts constitute the building blocks of our thinking and memory. The figments of imagination we experience are merely the products of ideas generated from our previous experiences, assembled together to form various new complex ideas (see 'Primary thoughts are *a priori* phenomena', above). It must be noted that *we are not capable of imagining anything totally new outside our own experience*. Our ideas are not as innovative as they may seem – even the most complex ideas entertained in our minds are merely the sensations experienced by us in our past. To illustrate this point let us run the following thought-experiment.

The Alien

Try to imagine an alien in your mind's eye – a grotesque alien of your weirdest imagination. You try to imagine this alien in such a way that you have not visualized such a thing in your past experience. And you must make sure that *all the individual parts of the alien* you visualize in your 'mental eye' are *entirely new and purely from your imagination*, but not from your past experience. You may try hard but you will not be able to do it. For example, you imagine that this alien has a face of a hornless goat, the body of a wooly mammoth, with hands like a seal's flippers, with a bi-lobed brain exposed in the center of its head shooting blue sparks of electrical activity, and eyes like glowing twin stars. It's really grotesque, is it not? Nobody could have envisioned such a weird thing – not even the iconic film director, Steven Spielberg! But if you closely examine this alien, all these parts of the imagined alien were only *a complex amalgamation of already known ideas*. Now we shall 'dissect' our alien to prove the point. A hornless goat is nothing new to the imagination – you have already known the form of a goat; you have imagined horns in your mind – and deleted them. In the same way, a wooly mammoth is possibly your idea from the pages of your science book; and perhaps the same thing with a seal with flat flippers; strange electronic circuits around the brain are shown abundantly in any sci-fi movies you watch (which can, in turn, be primarily experienced by watching lightning on a stormy night), and of course the shining stars are already there in the night sky. So what have you really imagined? Anything outside your past experience? Your alien is nothing new, but is only a product of several assorted old ideas. Thus, any imagination is merely a product of ideas we have drawn from our past experience (which, of course, are again made up of several immutable basic primary thoughts).

Dreams and Hallucinations

Dreams are also the products of memory – we cannot dream of anything which we have not experienced in the wakeful state. Even the oddest of hallucinations and delusions (experienced by the insane and the wise, alike) are the mere products of several assorted old ideas – we cannot imagine what we have not seen. The human sensory faculty has this limitation – *it cannot know what is unknown*.

Consider Other Examples

Every day we will be experiencing new ideas in the form of ***new things, new tastes, new music***, etc. – but all these complex ideas are concoctions of different sets of primary thoughts and ideas. For example, today we may experience a new dish with an exotic flavor – the flavor may be new to us, but each basic unit of taste and smell that make up the flavor is merely an immutable primary thought – the new flavor has arisen from a different combination of these old primary thoughts of taste and smell (which are already present in us, as explained above). Consider another example – we may listen to new piece of music; the music is certainly new, but all the *notes* are nothing but our good old primary thoughts of sound (see 'Examples of primary thoughts and ideas, above) – we 'weave' new music by using combinations of several primary thoughts of sound.

We may start imagining for ourselves *anything* and see for ourselves the inevitability of incursion of past experiences into our present thinking – all are oldish ideas woven into new complex ideas!

Fiction is No More Fiction

Yet another thought-experiment – you try to imagine a fictional character from your favorite comic. Say, whenever you imagine the character of *Superman* in your mind's eye, you may automatically get an image of the iconic actor Christopher Reeve from the movie. Or maybe you have in your mind the picture of Superman from your comic books. Or, you may imagine your Superman to resemble one particularly muscular and handsome man you know from your personal life. But you cannot 'build' a totally new Superman outside your own experience – that is simply not possible.

However, this phenomenon of the non-innovative nature of human thought is further discussed in Chapter 8 to illustrate this feature in a different light.

PRIMARY THOUGHTS ARE INHERITED

In this section it is demonstrated that primary thoughts are inherited. Genes play an important role in the formation of primary thoughts.

Explanation

What differentiates one individual from another individual? What makes us different from our neighbors? For that matter, what makes us different from a *chimpanzee*? Obviously it is the ***genes*** that we are made up of which make us different. Genes are nothing but small molecular machines that synthesize different kinds of proteins characteristic of a species, and consequently, certain ***proteins*** that make up our cells make us different from each other. How do these proteins make us differ from each other *physically*? Though all cells in each of us look the same at a gross level, their architecture and arrangement are slightly different from individual to individual so that we become different in so many physical characteristics. Thus, all the physical characteristics, such as *facial features, fingerprints,* our *HLA typing,* etc., are highly individual and vary from person to person. Even identical twins are genetically different from each other.

Not only are these physical characteristics distinctly individual, but even ***behavioral characteristics*** are different from one individual to the other. It is known that the way we talk (e.g., *voice-recognition patterns*) or the mannerisms we exhibit, or the way we walk (e.g., *gait-patterns*), are all different from person to person. Finally, we may say that each individual has his or her own ***emotional (temperamental) characteristics*** – each individual *thinks and acts differently* from others in a given situation. We have also seen that all primary thoughts are highly individual in perception, and they differ slightly between individuals (see 'Note' in 'Primary thoughts are non-comparative', above). We can now conclude that:

> The set of proteins we inherit are responsible for physical, behavioral and emotional characteristics that we inherit from our parents.

It has long been known that proteins are involved in the mechanism of learning and memory in animals (including human beings). It is experimentally established that protein synthesis in the brain is accelerated during learning. It has also been experimentally proved that if protein synthesis is blocked in neurons, the formation of *long-term memory* is abolished (Barrett et al., 2012, p. 285; Ch. 8, p. 159). Furthermore, it is shown in Chapter 7 (p. 133) that molecular gadgets are chiefly made up of proteins. Finally, based on the above information, it can be deduced that proteins are needed to generate primary thoughts, and different sets of proteins give us different kinds of thoughts (see Ch. 7, p. 140).

We have a 'bank' of molecular gadgets in our neurons that are made up chiefly of proteins, and which can generate primary thoughts. The neurons containing these molecular gadgets are present at the time of birth – i.e., they are already lodged in the basic structure of an individual's mind. If such is not the case, we could not have received and deciphered the sensation of blue light and converted it into a primary thought, as shown in the thought-experiment at the beginning of this chapter. Our *inability* to decipher ultraviolet rays (as discussed above) is also due to the *lack* of these protein molecular gadgets. To sum up, it can be said that our genome (the entire genetic information in the form of DNA present in a cell) has the inherent ability to produce all the sets of proteins necessary for the generation of all types of primary thoughts concerned with the individual (Ch. 7, p. 140).

Psychologists at the University of Liverpool, UK, have found that children as young as 2 years *have an understanding of complex grammar* even before they have learned to speak in full sentences (Noble et al., 2011). Does this not mean that human beings are fundamentally grammar-oriented? And, does this not also mean that the molecular gadgets responsible for *basic rules* of grammar and syntax are already lodged somewhere in the human neurons? We have to wait till Chapter 8 for a full explanation of this phenomenon.

So far we have studied the properties of primary thoughts and ideas. Now in Chapter 5 we will see in what type of neurons these primary thoughts are generated in the brain, and how they assemble to form ideas.

CHAPTER

5

The Perceptual Neurons

OVERVIEW

Chapter 4 discussed the characteristic features of primary thoughts, and also ascertained that ideas are formed by an assembly of primary thoughts. We have also seen that the generation of these primary thoughts needs some sort of molecular gadgets in the neurons. Now we will see what type of neurons house these molecular gadgets, and where these neurons are located in the human central nervous system.

Before starting with the study, let us first put forward the following three questions:

1. Are there any special types of sensory neurons that are designed to convert the peripheral signals into primary thoughts?
2. If there are such special neurons, where are they located in the CNS?
3. What evidence do we have to ascertain their presence?

The answer to the first two questions is given within the first three main headings of this chapter, wherein a brief study is undertaken to understand the neural pathways involved in the transmission of peripheral sensory signals to the CNS. It is shown that different types of sensations (such as vision, touch, pain, etc.) are perceived at different levels in the CNS – some at the cortex level, some at the thalamus level, and some even lower. This analysis is shown to have an important bearing in identifying the exact location of the special type of neurons, called *perceptual neurons*, which are responsible for converting peripheral sensations into primary thoughts.

The third question is answered below, under the heading, 'Locating perceptual neurons in the cortex'. Here the weight of scientific evidence that supports the existence of such special neurons in the human brain is presented. Now we will proceed with a brief account of the pathways of perception.

The Biology of Thought.
DOI: http://dx.doi.org/10.1016/B978-0-12-800900-0.00005-1

PATHWAYS OF PERCEPTION

All **stimuli** we receive from the external world reach the receptors, where they are converted into **sensations** that reach the CNS, and there they are converted into **perceptions** or **primary thoughts** (see Ch. 3, 'How is memory initiated?' & Fig. 3.1). For example, when we look at the color blue, the eyes receive the sensation of blue and the brain perceives the thought of the color blue. It is known that all the **special senses** are chiefly perceived at the cortex level – vision in the visual cortex, sound in the auditory cortex, and smell in the olfactory cortex, and so on. However, it is now well established that the **general senses**, such as touch, pain, temperature, etc., are perceived at different levels in the CNS (Guyton & Hall, 2008, p. 557–558), as discussed below. We will take **touch** and **pain** as examples of general senses and see at what levels these sensations are perceived in the CNS.

Before going into the neural pathways involved in the transmission of touch and pain sensations, the reader is advised to go back to the start of Chapter 1 and review the general plan of the nervous system. Here we will briefly see how exactly these signals travel, what pathways they take, and where they are converted into primary thoughts. The concept of perceptual neurons is introduced in this chapter.

Touch Sensation

Touch (tactile) sensation can be grossly divided into two categories – *fine touch* and *coarse touch*. Fine touch is a highly refined sensation developed fully in human beings and is especially concentrated at sensitive areas of skin (e.g., finger tips). It is perceived and analyzed at the cortex level. Coarse (or crude) touch is an ill-defined sensation, spread throughout the body, and is perceived *below* the cortical level, as we will see below (Fig. 5.1).

Fine Touch

The receptors for fine touch are many, ranging from *free-nerve endings* to *Meissner's corpuscles*, and they are designed to receive the mechanical stimulation of touch and transduce (i.e. convert) these stimuli into electrical signals. These signals then pass along the nerves (see Ch. 2, p. 32) and reach the dorsal root ganglia (Fig. 5.1). Axons emerge from the ***dorsal root neurons***, enter the spinal cord and pass in the *dorsal column*; these nerve fibers constitute the ***first-order system***. From here the axons pass through the *medial lemniscal system* of the medulla oblongata and end on the neurons of the thalamus; this constitutes the ***second-order system***. There is evidence that these first- and second-order systems do not change the quality of fine touch sensation, but have the important function of abolishing their transmission, if it becomes necessary. They

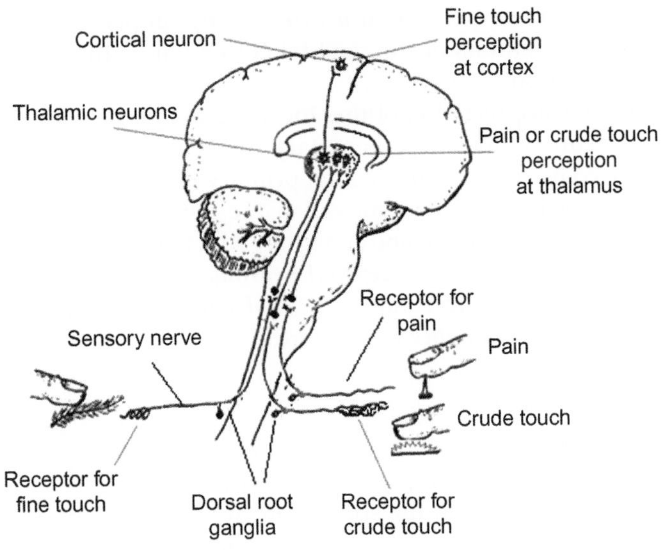

FIGURE 5.1 Perceptual neurons at various levels.

act as stop-gaps. In fact, the thalamus (see Ch. 1, 'The diencephalon') is a big 'yes-or-no' center, in the sense that it can either transmit the signals or keep them in check; for example, we do not always feel the clothes we are wearing or the perceive the sensation of touch on our backs when we sit on a chair, though constant touch signals are being sent from our skin surface – these 'routine' sensations are abolished by the thalamic centers. If such were not the case, life would become horrible with the multitude of sensations continually bombarding the brain and cluttering the mind!

From the thalamic neurons, axons project up to the neurons of the cerebral cortex, which constitutes the ***third-order system*** (Guyton & Hall, 2008, p. 589). At the cortex the fine touch sensations are finally appreciated as conscious perception – and these perceptions constitute the formation of the ***primary thoughts of fine touch.***

Note: The above description is an over-simplification of the actual pathways that take place in the CNS. The neural course in fact is very complicated, but for the present purpose this discussion will suffice.

The Concept of Perceptual Neurons

For details see 'The perceptual neurons', in p. 98. From the above discussion it becomes evident that the signals of ***fine touch*** are ***not*** appreciated in the spinal neurons or the thalamic neurons, but are simply passed up to the cortex. In other words, these spinal and thalamic neurons are ***not*** designed to generate primary thoughts of fine touch. Hence, these neurons are called ***non-perceptual neurons***. They simply pass up the signals to reach the neurons of the cerebral cortex where the signals attain conscious perception.

The conscious appreciation of fine touch occurs in the cortex, because the cortical neurons house the necessary molecular gadgets, which are designed to convert the signals into primary thoughts of fine touch. Hence, these cortical neurons are called the ***perceptual neurons***. What these perceptual neurons are and how they function is discussed below.

Hence, for ***fine touch*** the ***perceptual neurons*** are located in the ***cortex***, and these perceptual neurons house the ***molecular gadgets*** to convert the signals into ***primary thoughts*** of fine touch.

Coarse (Crude) Touch

We will now look into the transmission of coarse touch (Fig. 5.1). The receptors for coarse touch are also of many types (e.g., *Ruffini's end-organs*). They transmit signals up to the spinal cord where they end on the neurons of the *anterolateral system*. These are the first-order neurons and they project to the second-order neurons present in the brainstem or the thalamus. There is considerable experimental evidence in animals that even if the cortex is destroyed, coarse touch is appreciated almost normally. There is also sufficient clinical evidence in human beings to support this view. (Not only coarse touch, but the more basic sensations such as pain and temperature are also appreciated almost normally in the absence of cortex, as described below). This means that the cortex is not particularly essential for the appreciation of coarse touch, and it is perceived at the level of ***thalamus***.

Hence, for ***coarse touch*** the ***perceptual neurons*** are located in the ***thalamus*** (*not* in the cortex, as in the case of fine touch), and these perceptual neurons house the necessary ***molecular gadgets*** to convert the signals into ***primary thoughts*** of coarse touch.

Note: However, it must be considered that coarse touch in the case of humans is transmitted to the cortex for intelligent analysis, but these perceptions are no longer primary thoughts, as they are the result of an assembly of several primary thoughts – and thus they are *ideas* – as shown in 'Simple ideas and complex ideas', below.

Pain Sensation

Pain is a more basic sensation and constitutes an important warning signal. Perhaps pain, coarse touch, heat, cold and other such basic sensations that are vital for survival are *not dependent* on a well-developed cortex for their perception. There is evidence that pain is perceived normally even in decorticated (=cortex removed) animals (Guyton, 2008,

p. 600–601). This naturally means that pain perception should occur below the level of cortex (Fig. 5.1), and it is known that it is perceived in the thalamus, or in the brainstem, or perhaps even in the spinal cord (Guyton & Hall, 2008, p. 558 & 601, see discussion below).

The pain receptors are the *'free-nerve endings'*, which convert noxious (harmful) stimuli into electrical signals that are sent to the spinal cord, much like those of crude touch signals. Pain sensation travels in two major pathways – the ***fast-sharp pain*** *pathway* and the ***slow-chronic*** *pathway* (Guyton & Hall, 2008, p. 600). The fast-sharp pain is chiefly transmitted by the *neo-spinothalamic tract* (neo = new) and these sensations are perceived in the somatosensory cortex. This pain modality is very precise, and its neurotransmitter is *glutamate* (see Ch. 2, p. 51). The slow-chronic pain is sent up by the *paleo-spinothalamic tract* (paleo = old), which is poorly localized, more primitive in nature, perceived in the thalamus (Fig. 5.1), and is probably mediated by the neurotransmitter *substance-P*.

Levels of Pain Perception

It can be shown that various kinds of pain can be perceived by the perceptual neurons located at different levels in the CNS, as described below:

- **Spinal cord**: the sensation of sharp pain, for example, reaches the ***spinal neurons*** and these signals are directly shunted to the motor neurons of the spinal cord (of course, via interneurons, see below) and this results in a reflex muscular action of the affected limb; this constitutes the ***reflex arc***. This shows that the spinal sensory neurons can act independently (Guyton & Hall, 2008, p. 558) in the case of acute pain. In other words, here the spinal neurons are the perceptual neurons. Hence, for some cases of acute pain the ***perceptual neurons*** are located in the ***spinal cord***; they can analyze the pain sensation and act on their own (without conscious cortical supervision).
- **Brainstem**: when a pin pricks the finger unawares, there is a sharp gasp, sweating, and a quickening of the pulse. The reason for these reflex phenomena is that certain sensations of acute pain are relayed to the ***brainstem*** neurons that mediate these autonomic reflexes. Here the brainstem neurons are the perceptual neurons, and act independently without cortical control (Guyton & Hall, 2008, p. 601). Hence, for some cases of sharp pain the ***perceptual neurons*** are located in the ***brainstem***; they can analyze the pain sensation on their own (without conscious cortical supervision).
- **Thalamus**: moderate degrees of pain are relayed to the ***thalamic*** neurons, where pain is appreciated with some discretion (Fig. 5.1), making it possible for us to *bear* the pain if it is moderate (for example, we can endure the pain of a heavy weight on our shoulders for some time) or *ignore* it, if mild (see 'Fine touch' above). Hence, for

some cases of moderate pain the ***perceptual neurons*** are located in the ***thalamus***; they can analyze the pain sensation on their own, and thus the ***primary thoughts*** for this pain are generated here.
- **Cortex**: pain signals may finally be relayed to the ***cortical*** neurons for a more elaborate analysis. The exact location of perceptual neurons in the cortex is discussed in 'Locating perceptual neurons in the cortex', below. Hence, for ***pain*** sensation the ***perceptual neurons*** may be located in the ***spinal cord***, the ***brainstem***, the ***thalamus*** or the ***cortex***, and these perceptual neurons house the necessary ***molecular gadgets*** to decipher the ***primary thoughts*** of pain at each of the above levels.

Now, we will see what these perceptual neurons are and what exactly they do.

PERCEPTUAL NEURONS

Perceptual neurons (Fig. 5.2) are specialized sensory neurons that are designed to convert the incoming sensory signals into primary thoughts. They are involved in the conscious appreciation of a sensory modality, thus they are where primary thoughts are generated. We have seen that in the case of fine touch sensation the first- and second-order neurons present in the spinal cord and thalamus *cannot* convert the sensory signals into primary thoughts, and hence they are called the *non-perceptual neurons* (see 'The concept of perceptual neurons', above). On the other hand, the

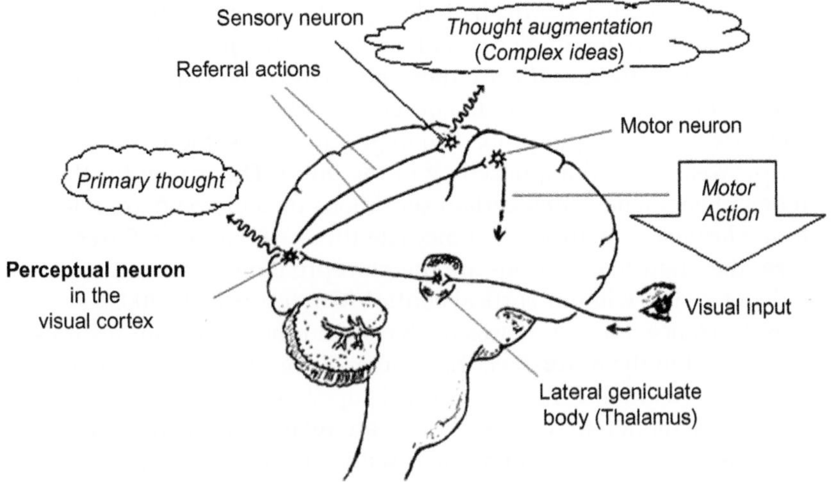

FIGURE 5.2 Various functions of perceptual neurons.

cortical (third-order) neurons convert signals of fine touch into primary thoughts, and these are called the *perceptual neurons*. These perceptual neurons lodge the molecular gadgets necessary to convert sensory signals into primary thoughts. These perceptual neurons have two important functions:

- The generation of primary thoughts
- Orchestrating referral actions.

The Generation of Primary Thoughts

The chief purpose of a perceptual neuron is the conversion of peripheral stimuli into primary thoughts (Fig. 5.2). In the above example, the sensation of fine touch is carried to the sensory cortical neurons, where the nerve impulses are converted into 'thoughts of fine touch'. However, the perception here is an ***abstract process*** – i.e., no obvious structural changes occur in the body in the process of conscious appreciation of sensations (see below). The following example makes the point clear.

When we look at a rose flower, feel its texture and smell it, we will receive the sensory signals pertaining to its color, its texture and its smell, and the perceptual neurons at various cortical areas convert these sensations into primary thoughts of color, touch and smell (see Fig. 3.6). These primary thoughts assemble to form various ***ideas*** of color, texture and smell to appreciate finally that we are looking at a rose flower (Ch. 9, p. 171). These primary thoughts give us only a *'feeling* of color', a *'feeling* of its texture' or a *'feeling* of smell', which together make up the mental image of a rose. These feelings are abstract in nature (Fig. 5.2), in the sense that they do not involve any gross physical change in the body (as seen in a motor process, which cause a physical or structural change such as muscular contraction or glandular secretion etc.; see 'Referral actions', below). If a student looks at a math problem on the blackboard he can solve the answer in his mind; this is an abstract mental 'feeling' but not a motor process (only when he writes the answer in his notebook does this become a motor action, as in 'Referral actions', below). This means that there is no gross physical change in this process of generation of primary thoughts. But we know that these primary thoughts make up most of our mind (Ch. 9, p. 168).

However, it should be noted that certain molecular changes *do* occur in the perceptual neurons, which will lead to the generation of primary thoughts. Please see Chapters 6–8 for a discussion of where these changes occur and what these mechanisms are. For now, this chapter will only identify the exact location of perceptual neurons.

Referral Actions

Apart from the generation of primary thoughts the perceptual neurons have the following three referral functions (Fig. 5.2).

1. The primary thoughts that are generated at the perceptual neurons of the sensory cortex may be relayed to the motor cortex for a proper ***motor action*** (Fig. 5.2). Here the perceptual neurons refer the information to the motor neurons (via the interneurons, see below). This may result in either contraction or relaxation of muscle fibers of the skeletal muscle or smooth muscle, or changes in secretion of glands, or changes in the function of organs such as heart, lungs etc. For example, if we look at a rose and want to pick it, the perceptual neurons not only appreciate the beauty of the rose, but they refer the information to the motor neurons for sending appropriate messages to the muscles of hand to pick the rose. Or sometimes, the perceptual neurons may (directly or indirectly) refer the stimuli to the smooth muscles of the blood vessels, intestine, skin, iris etc., which may constrict or dilate to show the desired response – for example, a tickling touch sensation may raise goose bumps by making the smooth muscles of hair-follicles in the skin contract. Sometimes the perceptual neurons may send information to the glands to pour out their secretions – for instance, when we taste a sour lemon, our mouths release saliva. All these effects are overt physical manifestations.
2. The perceptual neurons may also refer the primary thoughts to some other perceptual neurons thereby leading to the phenomenon that we may call ***thought-augmentation*** (Fig. 5.2). However, all these are also abstract processes, and are examples of complex ideas (see below). This transmission of signals from one perceptual neuron to another can be very extensive, so that a single stimulus may lead to the formation of a variety of ideas. For example, when we look at a single rose flower from a distance, we may be able to automatically think of its aroma, its texture, its associated thorny nature and a vast number of other past memories associated with roses. Human intelligence is largely a product of this thought-augmentation, which in turn is dependent on the long-term memory (LTM) of that individual (see Ch. 8, p. 160).
3. Finally, sometimes the information 'dies down' without getting conscious appreciation (such as the dinghy boat in Ch. 3). This is the ***incognisant state***. We do not know much about this phenomenon.

These referral actions are more diverse and complex, which are mediated by several special intermediary neurons called ***interneurons***. The interneurons deserve a special mention here in this context.

Interneurons

Conventionally, sensory neurons and motor neurons together are called ***principal neurons***, and in between these principal cells are the connecting neurons called ***interneurons*** (see Ch. 1, p. 5 & Fig. 2.1). These interneurons far outnumber the principal neurons in the CNS of human beings. It has been already shown that the sensory neurons will not connect directly with the motor neurons, because each and every time a stimulus passes from the sensory neuron to the motor neuron, the motor neuron *has* to respond without protest and it becomes like a puppet – every single pull at the string results in an awkward action (Ch. 2, p. 34). There must be a way by which the actions of motor neurons are moderated, and here comes the versatile play of interneurons. These interneurons act like a ***switching mechanism*** in the circuit – some interneurons may either pass the messages to the motor neurons (*facilitatory interneurons*), and some may check the messages (*inhibitory interneurons*).

However, a majority of the interneurons appear to be inhibitory, so that they can abolish incoming sensations. We have seen that we are oblivious to most sensations that bombard our senses from the external world, unless we pay specific attention to them. For instance, we do not hear the ticking of a wall clock all the time, nor do we feel the constant touch sensations of our clothes (Ch. 1, p. 6). Current research shows that, while most of the interneurons target principal cells, some target other interneurons preferentially – they are called *interneuron-selective interneurons* (IS cells) (Freund & Kali, 2008) – so that the messages have to pass from one interneuron to the other before reaching the principal neuron. The IS cells act on the motor neurons through a chain of interneurons, and this appears to increase the versatility of our actions.

The role of facilitatory interneurons is also significant. These interneurons augment sensations by spreading the sensory stimulus to many other principal neurons to show a more pronounced effect; for example, the mere sight of our favorite dish (visual input) may leave us salivating (autonomic response) and may even elevate our mood (emotional response).

Finally, it can be said that the most accomplished of these interneurons appear to *generate primary thoughts themselves* (and thus may act like perceptual neurons themselves). It is more likely, at least in human beings and some higher animals, that these specialized interneurons do most of the thinking. Consequently, we may presume that these specialized interneurons acquire the knowledge of the external world and store it in LTM for future retrieval. This confers human beings their indubitable capacity for memory, learning, innovation, intuition, creativity, and other such higher mental faculties. It is no wonder these interneurons outnumber the principal neurons in the human brain.

SIMPLE IDEAS AND COMPLEX IDEAS

We also ascertained in Chapter 4 that primary thoughts are the fundamental units of thought, and the ideas are formed as an assembly of several primary thoughts. Now we will see what simple ideas and complex ideas are, and how they form.

- ***Simple ideas*** result from the assembly of a limited number of primary thoughts generated usually within the *same cortical area;* for example, perception of the color blue is a primary thought, whereas the characteristics of the color blue, such as its hue, brightness, saturation, etc., are simple ideas (Fig. 5.3).
- ***Complex ideas*** result from the assembly of many primary thoughts, generated usually from *different cortical areas;* for example, if we can give the color blue the name 'blue', it becomes a complex idea because the primary thought of blue color must be associated with other primary thoughts of the auditory signals of 'blue', which is part of semantic memory, the centers of which are located at a distant cortical area (the temporal cortex; Fig. 5.3). Likewise, the ideas of blue sky or blue wallboard are highly complex ideas.

We will now take the human cortex as an example and see how primary thoughts assemble to form simple and complex ideas. Before going into this, let us examine the current concepts of how information is processed in the human *sensory cortex.*

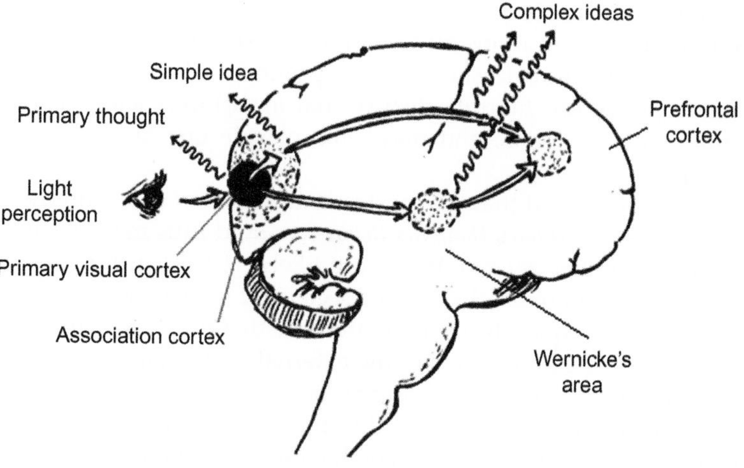

FIGURE 5.3 Primary and association areas: primary thoughts and ideas.

The Human Sensory Cortex

We have seen that general senses, such as touch, pain, temperature, pressure, vibration and proprioception, are appreciated in *somatosensory cortex* (post-central gyrus); special senses are perceived in the *occipital cortex* (vision), *temporal cortex* (hearing and vestibular), *olfactory cortex* (smell) and *insula* (taste) (see Ch. 1, p. 9). All these areas put together constitute the sensory cortex. It has been shown that in all these sensory areas the cortex is typically arranged in two parts (Guyton & Hall, 2008, p. 641) (Fig. 5.3):

- A central ***primary area*** and
- Peripheral ***association areas***.

Our current knowledge of ***color vision perception*** is fairly well advanced; hence we take color vision as an example to study how the sensory signals are processed in the cortex. We have seen that the sensation of light reaches the lateral geniculate body (a part of thalamus) and from there it is referred to the visual cortex for appreciation of color perception (see Ch. 3, p. 55 for details of this pathway). There is evidence that these visual sensations first reach the ***primary visual cortex*** (**V-1**; Fig. 5.3) where the primary colors (*red, green* and *blue*) are deciphered (more details in 'Locating perceptual neurons in the cortex', below). From the primary cortex the information is referred to the surrounding ***visual association areas*** where the other features of the color, such as its hue, brightness, saturation, etc., are deciphered. There are about 12 different visual association areas surrounding the V-1 (Guyton & Hall, 2008, p. 642). Perhaps the various secondary colors and shades of color such as yellow, magenta, purple, brown, etc., along with *forms* and *shapes* of the external objects, are deciphered in these association areas. And then this processed information from the visual cortex reaches the ***temporal lobe***, where a *name* is assigned to these perceptions of color – for example, we can now identify that this color is 'blue' or 'red' or 'brown', or whatever (see semantic memory, Ch. 3, p. 61). This visual information may again be referred to other ***distant cortical regions*** such as frontal and prefrontal cortex (Fig. 5.3), and this enables us to associate these colors with other ideas to make these ideas much more complex; for example, we can now associate the color blue with blueness of the sky, or an expansive wallboard, as seen in Chapter 4.

Generation of Simple Ideas and Complex Ideas

Now, if we analyze the above information and apply the concept of primary thoughts and ideas, this leads us to the following conclusion (Fig. 5.3 & Table 5.1):

- ***Primary thoughts*** of color are generated in the neurons of the ***primary visual cortex***; hence the primary colors red, green and blue

TABLE 5.1 Formation of Primary Thoughts, Simple Ideas and Complex Ideas

Stimulus	Stimuli from the external world	Light source taken as example	
Sensation	Stimuli converted into sensations at the receptors	Cones in retina	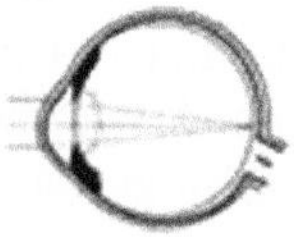
Perception	Sensations converted into primary thoughts at brain centers	Primary visual cortex (V1)	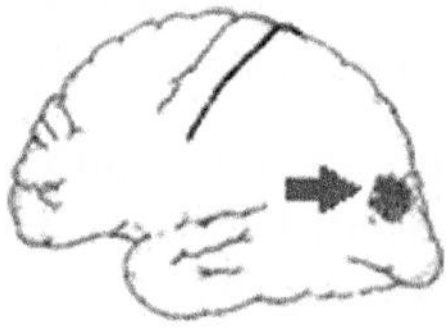
Assembly	Several primary thoughts assemble to form an idea	Neuronal circuits in visual association areas	
Association	Several simple ideas associate to form complex idea	Several neural circuits involving distant cortices	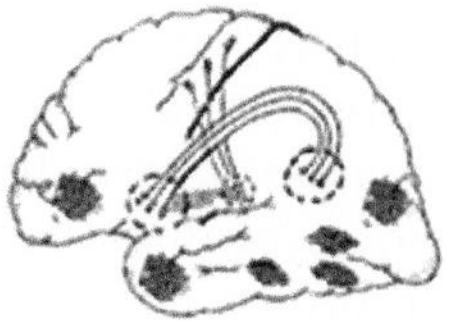

are primary thoughts. Therefore it can be said that the ***perceptual neurons*** for color vision are located in the primary visual cortex (V1). The exact location of perceptual neurons is shown below.

- ***Simple ideas***: these primary thoughts are then referred to the surrounding ***association areas*** where the secondary colors and other properties of color are appreciated. Hence, secondary colors such as purple, brown, orange, etc., are simple ideas; and color properties such as hue or brightness are also simple ideas.
- ***Complex ideas***: these simple ideas may be referred to other ***distant cortices***, such as the temporal lobe or frontal lobe and are converted into complex ideas; hence the *naming* of a color is a complex idea; and the appreciation of related ideas such as sky, wallboard, etc., are highly complex ideas.

Information Processing in Other Cortical Areas

In the above sections we have taken visual cortex as an example and studied how the visual information is processed. It is shown that in all other sensory areas the processing of information is somewhat similar. As an example, the ***primary auditory cortex*** (Area 41, Ch. 1, p. 12) first receives the sound signals. Thus, it can be said that ***perceptual neurons*** concerned with the generation of primary thoughts of sound are located in the primary auditory cortex and these neurons are concerned with deciphering basic units of sound (e.g., the perception of '***note***' in a piece of music, Ch. 4, p. 84). It is known that injury to this primary area will result in failure of perception of even simple sounds.

The ***auditory association cortex*** receives signals from the primary auditory cortex (Guyton & Hall, 2008, p. 658). Auditory association areas are concerned with the formation of ***simple ideas*** by ascribing properties of sound, such as its frequency, amplitude, intensity, etc. ***Complex ideas*** are formed by an assembly of different simple ideas (e.g., the whole of the 'raga' and its meaning). Though lesions of the association areas will not abolish the sound perception, it interferes with their understanding. The most important association area of the auditory cortex is Wernicke's area (Ch. 3, p. 72), and lesions in this area can lead to loss of *interpretation of meaning* of sounds heard (Barrett et al., 2012, p. 293). The auditory cortex is also connected to distant areas such as the frontal lobe, parietal lobe, limbic system, thalamus, brainstem etc., which enable us to interpret sound signals in a variety of ways – like the famous Pavlov's dog, which upon hearing the sound signal releases acid secretion in its stomach!

General sensations such as touch, pain, pressure, etc., are interpreted in the somatosensory cortex by a somewhat similar mechanism. The somatosensory cortex has two sensory regions – a ***primary sensory area,*** also called **S-I** (Areas 1, 2, 3) and an ***association area,*** also called **S-II** (posterior parietal areas) (Ch. 1, p. 12). All the general sensations reach the S-I first, where they generate primary thoughts. The S-II receives signals from the S-I and plays an important role in *deciphering the meaning* of sensory input along with the ability to recognize complex objects (Guyton & Hall, 2008) – in other words, S-II assembles various primary thoughts to give rise to complex ideas.

Needless to say, our day-to-day experiences are composed of many complex ideas, in an organized array somewhere in the cortex. For example, the image of a rose flower we perceive by using our sense of vision, touch and smell is a highly complex idea comprising several groups of primary thoughts and ideas from different regions of the cerebral cortex (Table 5.1).

LOCATING PERCEPTUAL NEURONS IN THE CORTEX

What evidence is there that perceptual neurons exist in the CNS? We will now try to find a place where we can conclusively prove their existence, and that place is the human neocortex itself.

We will briefly study the microscopic anatomy of human cortex first. The human cortex is roughly 2–4 mm in thickness, and microscopically consists of six layers of neurons (see Ch. 1, p. 19). This type of six-layered organization is called ***neocortex*** (neo = new or advanced). Most of the human cortex consists of this type of neocortex (except in hippocampus, olfactory cortex and certain other limbic areas, where the cortex is three-layered and called *paleocortex*; paleo = old). These six layers are numbered from I to VI (see Fig. 1.7). Of these, ***layer-IV*** is very important, as we shall see below.

It is now well established that the neurons in the human cortex are grouped into small bundles that are arranged in ***vertical columns***, and each column is composed of all six layers (Guyton & Hall, 2008, p. 591; Fig. 5.4). Their appearance can be imagined to resemble little cones of layered chocolate-cake arranged in a honeycomb-like fashion over the surface of the cortex. Each vertical column has a diameter of 0.3–0.5 mm and contains a total of about 10 000 neurons. It is known that each column represents a single ***sensory modality*** – which means that each column represents a single modality such as pain, fine touch, crude touch, pressure, temperature, color, smell, and so on (each situated in their respective sensory cortices). Current research shows that these columns are further organized into smaller columns called ***minicolumns***. Each of these minicolumns has a diameter of 40–50 μm and contains about 80–120 neurons. It has been estimated that there are approximately two million columns in the human cortex (or about 2×10^8 minicolumns).

Transmission of Signals in the Neocortex

The functioning of the visual neocortex is well studied. Current research shows that the incoming signals *first* reach ***layer-IV*** of the primary visual cortex (V-1) (Guyton & Hall, 2008, p. 642; Fig. 5.4). Then the signals spread both 'up' (i.e., towards layer-I) and also 'down' (i.e., towards layer-VI). Evidence shows that the signals that arrive at layer-IV remain *'isolated'*, in the sense that they preferentially spread 'up' and 'down' only, but not sideways. This means that the signals which reach layer-IV neurons either ascend up towards layers I–III or descend down towards layers V–VI; but they *do not* usually spread laterally to the other vertical columns. Obviously, this means that there is minimal or no interference between the columns, thus effectively segregating the columns from each other, thereby improving the quality of the signal. It has also been demonstrated that the inhibitory interneurons prevent this side-to-side spread

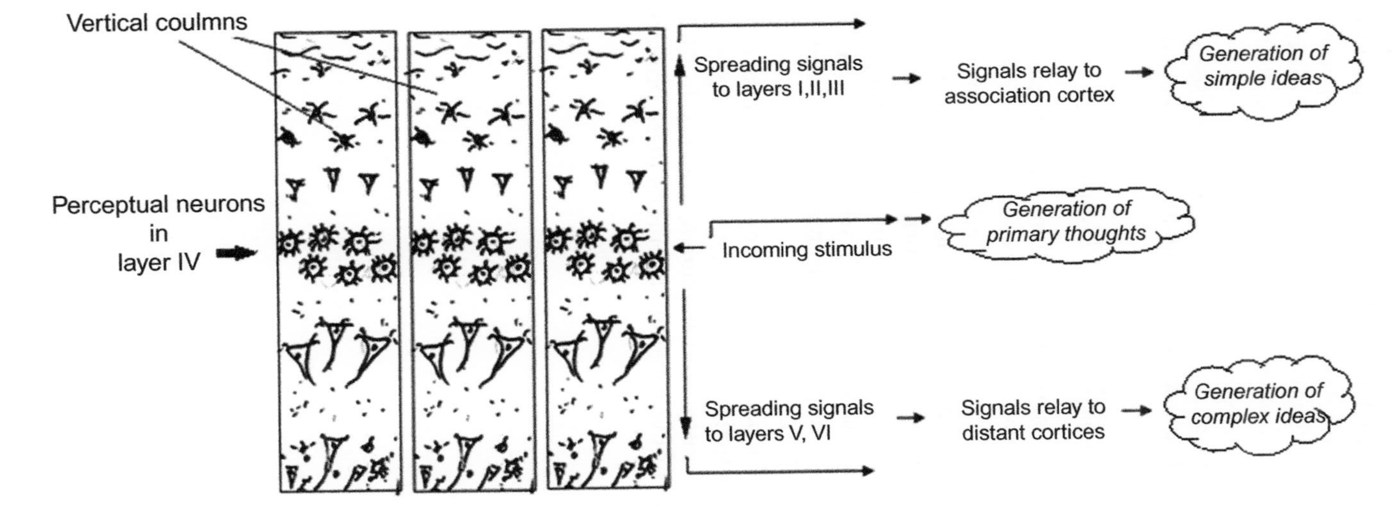

FIGURE 5.4 Vertical columns in neocortex generating simple and complex ideas.

of the signal among the columns. It is further established that the signals that spread up and reach layers I, II and III are relayed to the neurons of the surrounding ***visual association areas*** (Fig. 5.4), and the signals that spread down and reach layers V and VI are relayed to ***distant areas*** (such as the temporal lobe, frontal lobe, prefrontal cortex, etc.) (Guyton & Hall, 2008, p. 642).

Now we will utilize the above information to see how the primary thoughts, simple ideas and complex ideas are generated.

Level-IV Neurons as Perceptual Neurons

Thus, based on the above information, it can be said that the neurons in layer IV of the primary visual cortex constitute the perceptual neurons for vision, and these neurons generate the ***primary thoughts*** of vision (Fig. 5.4). From here the primary thoughts are transmitted to layers I, II and III, which are then relayed to the surrounding association areas where simple ideas are formed. The signals from the primary visual cortex may also be transmitted to layers V and VI, which are then relayed to the distant cortices, such as temporal lobe, frontal lobe etc. where complex ideas are generated (Fig. 5.4).

Having located the perceptual neurons in the human CNS, we will now proceed with the actual mechanism of generation of primary thoughts by the neurons, as discussed in Chapters 6, 7 and 8. In Chapter 6 we will see ***where*** exactly in the perceptual neurons these primary thoughts are generated.

CHAPTER

6

Dendrites and Primary Thoughts

OVERVIEW

Chapter 5 shows that there are perceptual neurons located at various levels in the central nervous system (CNS) and these convert external stimuli into primary thoughts. This chapter shows which neuronal structure is responsible for conversion of an external stimulus into a primary thought.

A neuron is like any other cell in its functions. However, neurons differ from other 'ordinary cells' in their electrical activity – and this capability is reflected in their immense ability to generate electrical signals and transmit them over vast distances in short periods of time. We will begin our discussion with a study of the electrical activity that is responsible for communication between the neurons, with an emphasis on the electrochemical events that happen at a synaptic junction. Before going into the discussion, the reader is advised to refresh his/her memory on the structure and physiology of neurons in Chapters 1 and 2, as this basic knowledge will be useful in the following discussion.

GENERAL OUTLINES

We have seen that all the external signals must be converted into electrical potentials before they are transmitted to the CNS because the neurons understand only the language of electricity (see Ch. 2, p. 33). It is also shown that the external signals pass up along the nerves to reach the perceptual neurons in the CNS, where they are converted into primary thoughts (Ch. 5).

Our task in this chapter is to know ***where exactly*** in a perceptual neuron does this process of conversion take place. To begin with we must first understand that communication in the CNS is chiefly ***electrochemical*** – because the neurotransmitters (NTs) are chemicals, and these chemicals

The Biology of Thought.
DOI: http://dx.doi.org/10.1016/B978-0-12-800900-0.00006-3

excite the neurons electrically (see Fig. 2.8). The most important structure for this electrochemical conversion is a synapse; hence we study the electrical events occurring in a neuron in the following order:

- Events at synapses
- Events at axons
- Events at dendrites.

In the last section in this chapter, we will make use of our knowledge of this electrical activity to arrive at a logical conclusion to pin-point the structure responsible for the generation of primary thoughts. Now let us study these structures and their electrical properties.

EVENTS AT SYNAPSES

It can be said that the whole study of the brain's functions is centered on these minute junctions between the neurons called ***synapses*** (Fig. 2.7). We will briefly study how an axon of one neuron transfers the signal to the next neuron through a synapse. We will take the example of light signals from the eye for our analysis.

Light signals are converted into electrical signals by the rods and cones in the retina, and sent via bipolar cells to the ganglion cells in the innermost layer of the retina (see Ch. 3, 'How is memory initiated?' for details). The axons of these ganglion cells are bundled up to form the optic nerve. These axons travel up and synapse with the neurons in the lateral geniculate body. Again, the axons arising from the neurons of the lateral geniculate body travel up to synapse with neurons of the visual cortex, where the perceptual neurons convert the signals into primary thoughts related to vision.

Axodendritic Synapses

The above discussion shows that at each stage in signal transmission, a synapse is involved. We have also seen in Chapter 2 that there are more than six types of synapses in the CNS, and the common types (in decreasing order) are *axodendritic synapse* (axon to dendrite), *axosomatic synapse* (axon to soma) and *axoaxonic synapse* (axon to another axon; see below).

Current research demonstrates that ***axodendritic synapses*** are exceedingly common in the CNS. It has been estimated that in the case of ***cortical neurons***, about **98%** of the synapses are axodendritic and only **2%** synapse with the soma (Barrett K. E. et al., 2012, p. 120). Thus, it can be said that axodendritic synapse is the most common type of synapse in the cortex (Fig. 6.1). It must be noted that even in the case of spinal neurons,

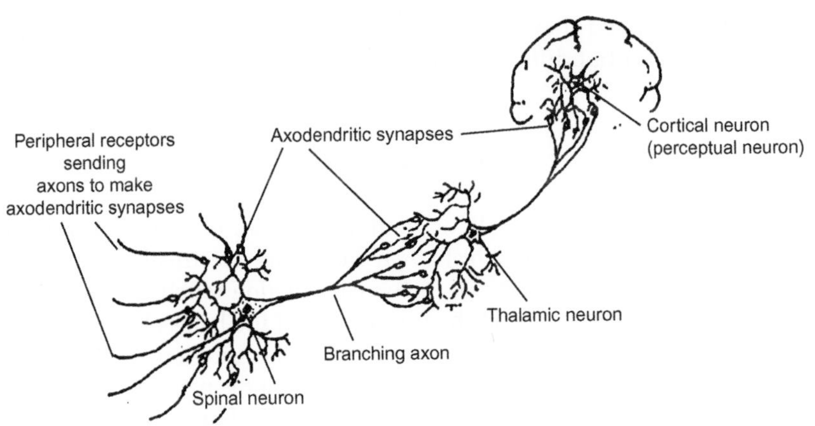

FIGURE 6.1 Axodendritic synapses.

axodendritic synapses are more common than axosomatic synapses but in an approximate ratio of 4:1.

We can now say that at each stage in the above visual pathway the ***axons*** of the transmitting neuron chiefly terminate on the ***dendrites*** of the next-stage neuron. Thus, the axons of the ganglion cells chiefly end up on the *dendrites* of the neurons in the lateral geniculate body and the axons from the lateral geniculate body end up on the *dendrites* of the cortical neurons. And finally, it can now be stated that axodendritic synapses are the chief mode of communication between the neurons in the CNS.

Unidirectionality

The chief property of a synapse is that there is an inherent unidirectionality in signal transmission across a synapse (Ch. 2, p. 42) – the signal passes from an axon to the dendrite, but not in the reverse direction (Barrett K. E. et al., 2012, p. 92). The chief purpose of this unidirectional flow of signals is that it allows the passage of signals towards a specific destination (Guyton A. C., 2008, p. 559). This feature can be translated into the following statement:

> The axons are the senders and the dendrites are the receivers.

Thus, a typical ***axodendritic synapse*** occurs between the terminal bulb of the axon of a sending (or transmitting) neuron and the dendrite of a receiving neuron. This implies that the signals from the axons are ***first*** transmitted to the dendrites of the receiving neurons. We will see in 'Events at dendrites', below, how the dendrites differ from the neurons and how specialized they are in receiving the signals (see also Table 2.2, to study the essential differences between axons and dendrites).

EVENTS AT AXONS

An external stimulus (e.g. light, sound or touch, etc.) excites the concerned peripheral receptor, which then incites an electrical signal called ***action potential*** (AP; see Ch. 2). In the case of fine touch the Meissner's corpuscles generate the APs, in the case of sound the hair-cells in the inner ear generate the APs, and in the case of vision the ganglion cells in retina generate the APs, and so on. These APs travel along the nerve fibers to reach their next-stage neurons (Ch. 5, p. 94). *All* these APs have a typical voltage of about 100mV. These next-stage neurons then get excited and generate another set of APs (which also have a voltage of 100mV), and these APs travel along their axons to reach the next-stage neurons (Fig. 6.1) – and this process continues until the signals reach their respective perceptual neurons.

An axon usually divides into several branches (Fig. 6.1) and the tips of these branches expand into specialized structures called *terminal bulbs,* which contain small vesicles (packets) loaded with chemicals, called NTs (see Ch. 2, p. 46 & Fig. 2.7). The APs reach the terminal bulbs and excite them electrically; they then release NTs into the *synaptic cleft* – which is a potential gap between the terminal bulb's membrane and the receiving neuron's membrane. The released NTs then act upon the *postsynaptic membrane* (the membrane of the receiving neuron, and in the case of axodendritic synapses dendritic membrane represents postsynaptic membrane). The action of NTs is to generate small *membrane potentials* (ranging from 1 to 50 mV, see below) on the dendritic membrane (see Ch. 2, p. 40 and 'Electrical activity at dendrites', below).

The following are the characteristic electrical properties of the axon:

1. The axon is designed to transmit only ***APs*** with a set amplitude of ***100mV*** (see Fig. 2.4); in other words, it cannot transmit signals with voltage potentials less than 100mV (compare dendrites, Table 2.2 and the section on 'Where in the Neuron?', below).
2. The characteristic feature of an axon is that it can be excited only if sufficient voltage is passed to it; conversely, if sufficient voltage is *not* supplied it fails to excite the axon at all. This property is called the ***all-or-none phenomenon.***
3. Another feature of APs is that, even though the axon divides into several branches the voltage potential passing through each branch is maintained at about 100mV only. This feature is due to the inherent property of the axon membrane itself (specifically due to the distribution of voltage-gated ion channels in its membrane).

'Labeled Line' Principle

After careful analysis of the above facts, we can arrive at an interesting observation. We have seen that all the APs share a common property: they have a *uniform* voltage potential of 100 mV. For example, the cones in the retina convert light into electrical signals and pass them as APs with a voltage potential of 100 mV, which travel up to the CNS. When excited the touch receptors in the skin also send APs with a voltage potential of 100 mV. Taste, smell, touch and all other receptors also only send their signals to the CNS by the way of APs at 100 mV.

This means that all the sensations are uniformly transmitted to the CNS with the same voltage – but they are perceived differently at the neuron level. Why do we perceive different perceptions, such as light, touch, taste, smell, etc., when the voltage potentials of all sensations are the same? The common explanation is that each axon terminates at a specific point or area in the CNS – this is called the ***labeled-line principle*** (Guyton 2008, p. 572–573); but the molecular-grid model (Ch. 7) offers a more clear and precise picture to explain this phenomenon, as we will see in subsequent chapters.

Compare this phenomenon with the phenomenon of ***phosphenes*** (as an explanation for why we see 'stars of light' when we rub our closed eyes) described in Chapter 4, and see the similarities between the labeled-line principle and phosphenes. Both of these phenomena stress an unimpeachable point that our sensations are exclusively perceived by neurons, and the receptors offer little or no contribution to our perception, except that they transmit signals.

Now we will study the characteristics of dendrites and see how they respond to stimulation, and how they differ radically from axons.

EVENTS AT DENDRITES

Dendrites (*dendron* = tree) are membranous tree-like projections arising from the body of the neuron, about 5–7 per neuron on average, and about 2 μm in length. They usually branch extensively, forming a dense canopy-like arborization called a ***dendritic tree*** around the neuron. The size of the dendritic tree is enormous compared to the soma – the size of the body of a neuron has been compared to a tennis ball placed in a large room and the size of dendritic tree to the large room itself, and consequently the dendritic membrane covers a much greater surface area than that of the soma. As shown above in 'Axodendritic synapses', axons preferentially terminate on the dendrites. And it is not surprising that about **75%** of the dendritic membrane of a typical neuron is said to participate in synaptic transmission!

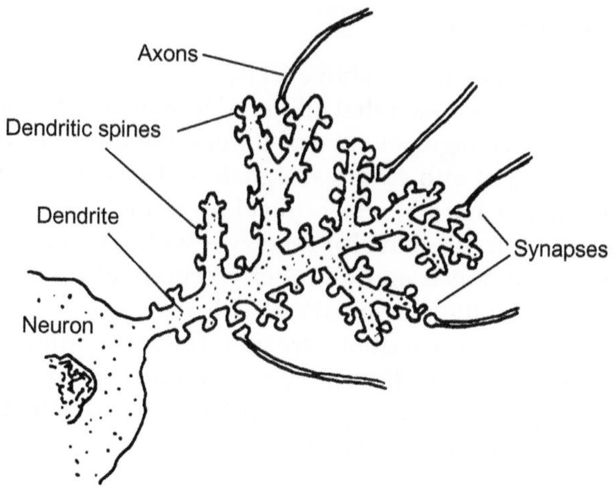

FIGURE 6.2 The dendritic spines.

Dendritic Spines

Dendrites show the presence of numerous spiky membranous protrusions called ***dendritic spines*** that deserve special mention because these are the structures that actually synapse with the axon's terminal bulbs (Fig. 6.2). A single dendrite has tens of thousands of dendritic spines (on average, 200000 dendritic spines per neuron!). Usually they have narrow stalks with bulb-like heads.

Dendritic spines are highly dynamic structures – changing their forms and synaptic connections with the passage of every signal to them. They may appear or disappear constantly with each given stimulus, giving the neuron its characteristic ***dendritic plasticity***, which was shown to be a part of synaptic plasticity (see below).

Synaptic plasticity is defined as the ability of neurons to bring about changes in the connections between neuronal networks in response to use or disuse. Synaptic plasticity is of two types: 1) mediated by the NTs, and 2) mediated by dendritic plasticity, both discussed in detail in Chapter 3. We are here concerned with the capacity of the dendritic tree to form new synapses or break old synapses in accordance with the stimuli that reach it from the axon terminals (Figs 6.3 & 3.9). The dendritic spines are capable of changing their forms rapidly in a matter of minutes or hours (Barrett et al. 2012, p. 125), which confers a great deal of dendritic plasticity to the neurons. This versatile property of quick response is especially true of neurons in the cortex, and is essential for rapid acquisition of memory and to enable learning. The reader may refer to Chapter 3, p. 70 to learn of other remarkable features of dendritic spines.

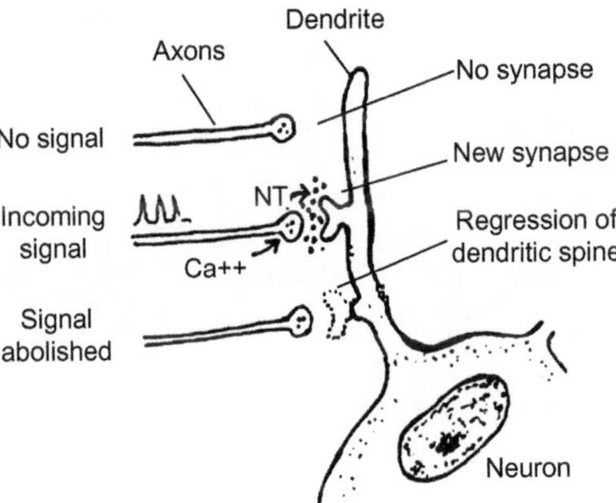

FIGURE 6.3 Dendritic plasticity.

Protein Synthesis and Dendrites

Protein synthesis is very prominent in the dendrites compared to axons. Current research shows that plenty of ribosomes (which synthesize new protein) migrate into the dendrites during signal transmission and learning. Furthermore, the cytoskeleton (which is essentially protein in nature) present in the dendrites is known to be poorly phosphorylated (= active form), in contrast to axon cytoskeleton, which is highly phosphorylated (= inactive form; see Table 2.2), indicating an active participation of cytoskeleton in the dendrites.

Pruning of Dendrites

Pruning of the dendrites is another feature of the dendritic tree that deserves a special mention. Pruning is the process of trimming unused dendrites in a dendritic tree during the development of the nervous system. Dendritic pruning is necessary to eliminate old and unused dendrites to facilitate the formation of new dendrites. This means that some of the old synapses are eliminated to allow the construction of new ones. In other words, this erases old memories to accommodate for new experiences. Hence pruning is also responsible for the acquisition of new memory and to facilitate learning.

ELECTRICAL ACTIVITY AT DENDRITES

The reader is advised to refer to Chapter 2, 'Signaling activity of neurons: a summary' for a quick review of electrical transmission in neurons. During signal transmission the axons of the transmitting neurons release

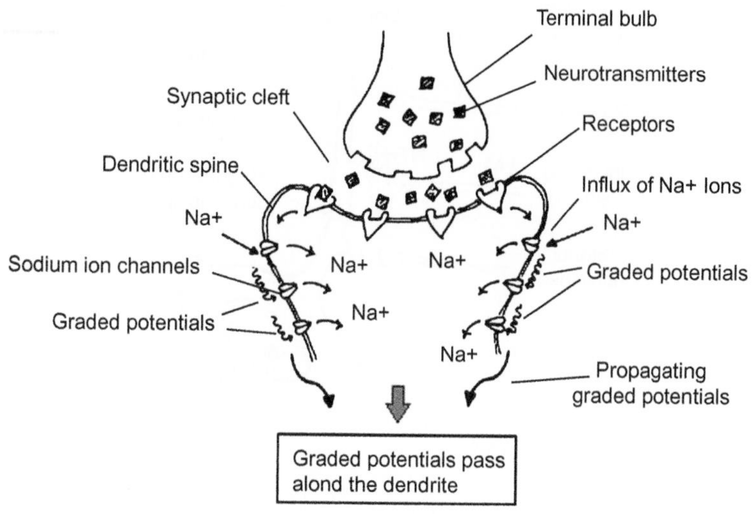

FIGURE 6.4 Generation of graded potentials.

NTs into the synaptic cleft. These NTs bind to the receptors in the dendritic membrane, which causes the ion channels to open up and allow ions to pass through the membrane (Fig. 6.4). All ions are electrically charged, and when they pass through the ion channels they generate small electrical currents across the dendritic membrane, which are then recorded as membrane potentials. This means that the *chemical energy* released at the synapse in the form of NTs is converted into *electrical energy* at the dendrites (see Fig. 2.8). These *electrotonic* potentials are called *graded potentials* (Fig. 6.4). Graded potentials are generated because of the presence of ligand-gated ion channels in the dendrites (see Ch. 2, p. 38 & Fig. 2.5). These dendritic potentials are very different from the APs generated by the axons – they are small voltage potentials with variable amplitudes of 1–50 mV and can travel only for short distances (whereas APs are large, with a voltage of 100 mV which can pass over long distances).

Graded potentials pass along the dendrites and reach the body of the neuron (soma) where they join with graded potentials from other dendrites. This process of the addition of potentials is called ***summation***. After summation these potentials travel down the neuron and collectively act on the axon hillock. The strength of these collected potentials is variable because numerous dendrites send graded potentials with variable voltages; if their collective strength crosses the threshold voltage required to excite the axon hillock then an *AP* is generated which is propagated along the axon (see Fig. 2.5). If the collective strength of these graded potentials is less than the threshold voltage, they cannot excite the axon hillock and they 'die down' subsequently. There is evidence that these dendritic potentials (but not APs) contribute to the formation of wave forms recorded in an *electroencephalogram* (EEG) (Barrett et al. 2012, p. 272).

What happens to the graded potentials before they die down? What is the result of this dendritic excitation? The role of graded potentials in the generation of primary thoughts becomes evident in Chapter 7.

Based on the above knowledge we will now proceed with the discussion of where in the neuron the primary thoughts are produced.

WHERE IN THE NEURON?

Now we will study which portion of the neuron generates the primary thoughts – is it the dendrites or soma or axons or synapses? Or is it the cytoplasm or nucleus or any other organelle?

Refuting 'The Whole Neuron Concept'

It is tempting to answer these questions by stating that the *whole neuron* is responsible for the generation of primary thoughts. In this age of science we know the intricate details of all the cytoplasmic structures of the cell and their functional significance (see 'Which portion of neuron?', below) – we know about the nucleus and its functions; we know the functional details of all organelles such as mitochondria, ribosomes, endoplasmic reticulum, etc. And in the preceding chapters we have seen the functional details of dendrites, axons and soma (as summarized below) – and it was shown that each of them has its own characteristic function.

A neuron is a cell and it has many cellular functions (apart from its electrical functions). A neuron (like any other cell) has an organization to use ***energy*** to perform its activities, such as respiration and metabolism; a neuron has an organization to maintain ***homeostasis***; a neuron has an organization to look after its ***defense*** mechanisms; a neuron has an organization to ***grow*** and ramify; and a neuron has an organization to undergo ***cell division***. Each neuron has a set of specific molecular mechanisms to deal with these specific tasks, and current science has a nearly complete knowledge about them.

Now we will make a point: much in the same way, a neuron also has an ***organization to generate primary thoughts***. It is our task here to find out what is this organization and where is it situated. Hence the 'whole neuron' concept is no longer suitable.

As we can see, there are two questions which must now be answered:

1. Which part of the neuron generates primary thought – is it the dendrites or soma or axon or synapse? (The section below answers this question.)
2. Which cell structure in the neuron takes part in their generation – is it the cytoplasm or nucleus or what else? ('Which portion of neuron?', below, answers this question.)

If we follow the logical deduction presented below, the answers to both of the above questions become apparent.

Dendrites Generate Primary Thoughts

The following analysis must convince us that the ***dendrites*** contain the ***molecular gadgets*** and they are the places where external stimuli are converted into primary thoughts.

First a summary of the specialities of dendrites is presented below:

1. About 98% of synapses in the cortex are axodendritic, as shown in 'Axodendritic synapses', above.
2. Dendrites are the receivers of signals in a typical axodendritic synapse (see 'Unidirectionality', above), and the dendrites sense the signals first.
3. The size of the dendritic tree is much more enormous than the soma ('Events at dendrites', above). This means that dendrites represent a much greater portion of the neuronal membrane than the soma, indicating their greater physiological significance.
4. It is shown that 75% of the dendritic membrane participates in synaptic formation. From the above four facts, it can be readily deduced that dendrites receive signals preferentially.
5. The dendrites are highly plastic in nature (see 'Dendritic spines', above). The dendritic spines are shown to participate in synaptic plasticity. The dendritic spines can either make new synaptic connections or break these connections with the passage of incoming signals to the dendritic tree. This type of plastic movement is possible only in dendrites (not in the axons or soma). This shows that dendrites are the most dynamic structures of a neuron.
6. Dendrites conduct graded potentials (see 'Electrical activity at dendrites', above). The voltage potentials generated at the dendrites are shown to be highly variable (1–50 mV). In the case of axons the action potentials are uniform. Hence modulation of electrical activity in the dendrites is easy.

7. The protein synthesizing capacity of the dendrites is enormous, which again-shows their active role in signal transmission. Animal experiments also show that protein synthesis is tremendously accelerated in the dendrites during learning. It must also be noted that in Chapter 7 it is clearly shown that molecular gadgets are chiefly protein in their composition.
8. Pruning is a unique property of the dendritic tree (see 'Dendritic spines', above).

All the above features show that dendrites are the sites of generation of primary thoughts.

Features of Soma

Look at the soma, which has none of the above features:

1. The size of the soma's membrane is comparatively small compared with that of the dendritic tree, as shown above.
2. Only 2–5% of axons synapse with the body in the case of cortical neurons.
3. There is virtually no plasticity in the body of the neuron.
4. Soma does not participate in pruning.

Features of Axon

Look at the axon, which also has none of the features:

1. Axons transmit only uniform action potentials with a voltage of 100 mV only. There is no variability in their transmission. It can be stated that axons are only transmitters (or senders) of signals, and they are not receivers.
2. The protein-synthesizing machinery in the axons is limited and cytoskeletal proteins are heavily phosphorylated (see 'Dendritic spines', above), which shows their relative inactivity.

Conclusion

Based on the details given above, it can be deduced that dendrites are capable of synthesizing and housing molecular gadgets, generating graded potentials, and converting them into primary thoughts. As the dendrites are found to be highly plastic in nature they are also responsible for abolishing some primary thoughts or allowing the generation of other primary thoughts. This feature is the cornerstone of the formation of human memory. This matter is taken up further in Chapter 8.

All these issues make the dendrites the most likely candidates for converting external signals into primary thoughts. Thus, we can make the following statement:

> The **dendrites** of the perceptual neurons are the structures which lodge the **molecular gadgets** necessary to convert sensory signals into **primary thoughts.**

Which Portion of Neuron?

The following analysis should convince the reader of the precise location of ***molecular gadgets*** and the exact place where external stimuli are converted into primary thoughts.

Is it the Cytoplasm?

We know the intricate molecular mechanisms of metabolic cycles that occur in the cytoplasm of a cell, and we know how the *mitochondria* (powerhouses of cells) give us energy. We have a fairly good understanding of the function of organelles such as *endoplasmic reticulum, Golgi apparatus, lysosomes, cytoskeleton,* etc. They are all useful in maintaining the energy needs or the structural integrity of a neuron. In the above discussion it is shown that the transmission of information between neurons is chiefly electrochemical, but none of these structures show any significant electrical (or ionic) activity. Voltage potentials have never been identified in these organelles. Thus it can be said that these organelles are ***not directly concerned with reception of a stimulus from the external environment*** – but they provide the neurons with the energy and support they need to perform electrical activity.

Hence, cytoplasm itself is an *unlikely* candidate for the presence of molecular gadgets.

Is it the Nucleus?

We also know the precise function of *DNA,* the *RNA,* the ribosomes, etc., and we know about the detailed structure of these nuclear structures. Genes are responsible for orchestrating all cellular events and modifying them as necessary. Genetic commands hold an important key for metabolism and for the efficient execution of other activities of the cell. Without genetic direction a cell cannot survive for long. Needless to say, genes also transmit all traits across generations. But none of these structures show any electrical activity – voltage potentials have not been identified in these structures. Our scientists have analyzed and synthesized DNA 'strand-by-strand' (see the research by Professor Craig Venter, who synthesized DNA in his lab!), but they have failed to notice any electrical activity capable of receiving APs from the external world. Thus, it can be said that genes ***do not receive sensations directly from the external environment***. (But it must be remembered that genes can ***modify*** the response to a signal by altering protein synthesis, and this is discussed in Ch. 7, p. 140).

Hence, the nucleus itself is an *unlikely* candidate for the presence of molecular gadgets.

Is it the Dendritic Membrane?

The most enigmatic structure of the neuron is its membrane. A neuron's membrane shows a lot of electrical activity in the form of various ***membrane potentials***. It must be noted that not only the neuronal membrane, but also cell membranes of all living cells show electrical activity (Hardin et al., 2012, p. 365). However, neurons specialize in generating versatile types of electrical potentials, as discussed in the preceding chapters. Dendritic membrane is also shown to be the structure which ***first*** receives ***signals from the external world***. These signals are deciphered at the membrane, and transmitted *into* the cell (by a *second messenger system*), so that the cytoplasmic organelles and nucleus can 'understand' them and respond suitably by synthesizing the appropriate proteins. Cell membranes are also ***capable of responding to external stimuli*** by orchestrating a florid display of membrane potentials. Thus, it can be said that dendritic membrane is the structure which receives signals from the external environment and converts them into primary thoughts. The cytoplasm and the genes support the membranes to execute this electrical function.

Energetics and Neuronal Membrane

There is another important aspect that emphasizes the importance of membrane in neuronal function. The human brain needs a huge amount of energy in the form of glucose and a lot of oxygen to function normally (see Ch. 2, 'Energetics'). Although the brain weighs only about 2% of the body's total weight, it uses about 20% of the body's total oxygen consumption and about 120 g of glucose per day (Hardin et al. 2012, p. 243). Where is this tremendous amount of energy expended? It has been estimated that up to two-thirds of a neuron's energy is spent in maintaining membrane polarity by activating the Na^+/K^+ pump. This shows that the neuronal membrane is the most active region of a neuron, and *utilizes large amounts of energy to convert external stimuli into primary thoughts.*

Although we have a fairly good understanding of the structure of biological membranes, we are far from precisely deciphering or 'mapping' the exact arrangement of their constituent lipids and proteins in the living membranes. The neuronal membrane will be further discussed in Chapter 7.

THE FINAL CONCLUSION

The **dendritic membrane** is the structure that houses the **molecular gadgets** necessary to convert sensory signals into **primary thoughts**.

The molecular gadgets hitherto described are called '***molecular grids***'. We will now proceed to Chapter 7, which examines a working model of thought generation by the neurons, called the ***molecular-grid model***.

CHAPTER

7

Molecular-Grid Model

OVERVIEW

The preceding chapter concluded that primary thoughts are generated by the molecular gadgets located in the membrane of the dendrites of perceptual neurons. These molecular gadgets are now called ***molecular grids***. Here we will answer the following questions:

1. What are these molecular grids composed of and how do they function?
2. What gives them the property of generating a variety of primary thoughts?

In the first three sections, we will study the intricate structure and function of the cell membrane and the participation of the membrane in signal transmission. In the next three sections, we will study a working model of these molecular gadgets, called the ***molecular-grid model***. We will examine their structure and function in converting signals into primary thoughts. In this chapter some solid physiological foundations are presented in support of the molecular model proposed. The genetic basis of primary thoughts will become obvious to the reader by the end of this chapter. Now we will proceed with the structure of the cell membrane.

CELL MEMBRANE

The essential feature of every cell is the presence of a membrane (= plasma membrane), which defines its boundaries. The cell membrane also limits the boundaries of cytoplasmic organelles such as mitochondria, nucleus, endoplasmic reticulum, Golgi bodies, lysosomes, etc. Not only do they form boundaries, they also form the external extensions of a cell, such as pseudopodia, cilia, flagella, dendrites, axons, etc., and all these structures have different functions. We will study the structure of a 'typical' cell membrane, i.e., the membrane present in *all* cells

The Biology of Thought.
DOI: http://dx.doi.org/10.1016/B978-0-12-800900-0.00007-5

(whether they be the brain cells, fat cells, skin cells, spermatozoa, plant cells or any other cells, because all of these have a universal arrangement in their membranes). However, it must be noted that the neuronal membrane has several special features that enable the neurons to generate and transmit electrical impulses with considerable speed over vast distances.

It is now established that membranes participate in various dynamic functions of a cell, for example homeostasis, cell-cell communication, cell division, cellular defense, etc. Although we know a lot about the structure of the cell membrane, we have some crucial voids in our knowledge of its exact arrangement in a *living* membrane. However, recent techniques such as *2-photon microscopy* and *optogenetics* can catch the cell membrane in 'live action'. Now we will study membrane structure and functions.

Structure of Cell Membrane

All cell membranes are composed of lipids (fats) arranged in two layers, i.e., in bilayer of lipids. Its outer layer, the ***exofacial layer***, is exposed to its surroundings; the inner layer, the ***cytofacial layer***, is exposed to the cytoplasm inside (Fig. 7.1). The membrane lipids are chiefly phospholipids, i.e., they have a phosphate *head* at one end, which is hydrophilic ('water-attractive'), and two lipid tails at the other end, which are hydrophobic ('water-repelling'; see below and Fig. 7.2a). All cells are surrounded by an aqueous medium and each cell also contains water (cytoplasm is 75–85% water!). Hence, the water-attractive phosphate heads need to face both outside and inside the cell (Fig. 7.2b). In effect, this produces a *thermodynamically stable bilayer* because the outer water-attractive surface of the membrane comes in contact with only extracellular fluid (ECF) and its dissolved ions; the inner water-attractive surface also comes in contact with cytoplasm with its dissolved ions, and the middle portion is composed of only water-repelling fatty chains and is 'protected' from the surrounding watery fluid. This water-repelling middle layer forms an effective barrier for the free transfer of water-soluble substances from one side to the other. However, for the life of a cell to continue, there must be a constant exchange of water, ions and other substances between the inside and outside of a cell through its membrane. How is this transfer possible? This transfer takes place because of a feature called ***selective permeability*** (see below), meaning that some substances are preferentially allowed to pass through it and some are not.

The currently accepted structure of the cell membrane is the ***fluid-mosaic model*** of Singer and Nicholson (modified by Unwin & Henderson; Hardin et al. 2012, p. 163). The 'fluid' part of the membrane is contributed by the *lipids (fats)* and the 'mosaic' part is contributed by various *membrane proteins*. The term 'fluid' here signifies that the lipids can move along the plane of the membrane freely from place to place and thus can carry the

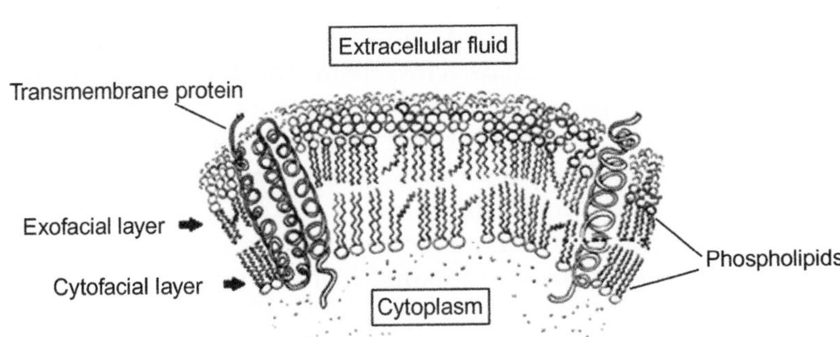

FIGURE 7.1 Structure of cell membrane.

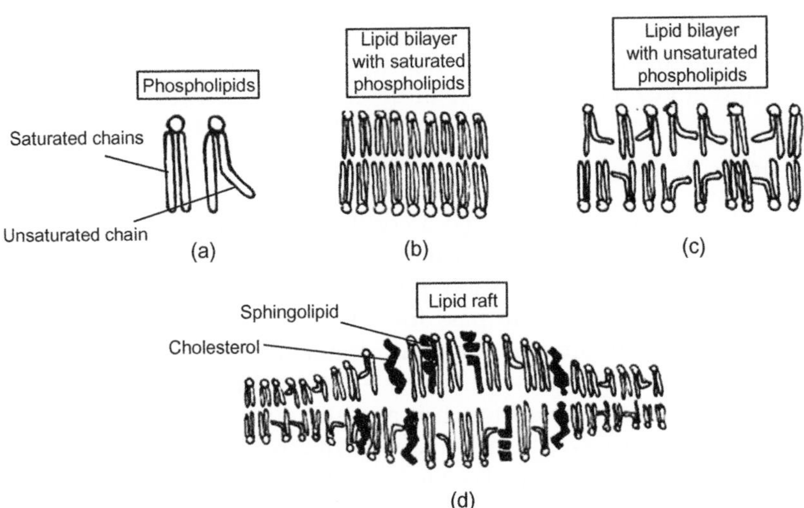

FIGURE 7.2 Lipids in the membrane.

membrane proteins along with them. Hence, the greater the fluidity of the membrane, the greater the mobility of membrane proteins. The versatile function of the cell membrane depends on its fluid nature, and this property is also responsible for the development of lipid-protein domains of the cell membrane, as we shall see shortly.

The majority of fats in the cell membrane are ***phospholipids***, which are composed of a *phosphate head* and two *fatty acid chains* (Fig. 7.2a). Fatty acids are of two types – *saturated* and *unsaturated*. Saturated fatty acids have straight hydrocarbon chains and so they are *compactly* packed together (Fig. 7.2b). This sort of tight packing makes the phospholipids 'immobile' and hence the membrane becomes merely 'solid sheets' of fat, which will not promote the versatile functions of membranes. However, membranes need to be more 'fluid' in order to allow free movement of proteins and certain special lipids across them. This is achieved by inserting a few *unsaturated fatty acids* into these sheets of saturated fats – the chains of unsaturated fats are bent and so *cannot* be tightly packed into solid sheets (Fig. 7.2c). Now, you can see that the membrane has become more fluid and can perform its functions.

However, note that a membrane must remain 'solid' at a few localized regions and the rest of the surrounding membrane must remain fluid (see 'The lipid component', below). This fluid–solid phasic transition is necessary for adaptation to surrounding conditions and for performance of other functions. This is enacted by inserting a few more special lipids into these layers of fat. The leader among these special fats is *cholesterol* (Fig. 7.2d). Cholesterol can behave uniquely – it can make membranes more solid or more fluid by virtue of its special bonding properties. In fact, cholesterol makes the membrane *less* fluid at high temperatures and *more* fluid at low temperatures (Hardin et al. 2012, p. 171).

Selective Permeability

As noted above, all cells need a constant transfer of water, ions and other substances across their membrane in order to survive. This passage across the membrane is only possible if it has ***pores*** in its structure that allow the substances to pass to-and-fro across the cell membrane. However, it must be stressed that cells cannot allow all substances across their membrane pores indiscriminately – the membrane has to allow only certain substances to pass though it and should block certain others. If such is not the case it would incompatible with the cell's life. This essential feature of the cell membrane is called ***selective permeability.***

This versatile feature is possible because of the presence of proteins guarding the pores. These are specialized proteins called ***transmembrane proteins*** (Figs 7.1 & 7.3) that span the entire thickness of the membrane (in contrast, some membrane proteins do not pass the entire thickness of the membrane but are limited to either the exofacial layer or the cytofacial layer). Transmembrane proteins have channels within them that selectively allow the substances to pass through. They have long and coiled segments of non-polar amino acids in their structures so that they can accommodate themselves in the non-polar lipid bilayer. The transmembrane proteins also have segments that protrude to the outside or inside of the cell, and these segments are chiefly made up of polar amino acids.

Types of Transport

Now we will briefly see how this transfer of substances takes place across the membranes (Hardin et al. 2012, pp. 197–216). Only a few simple

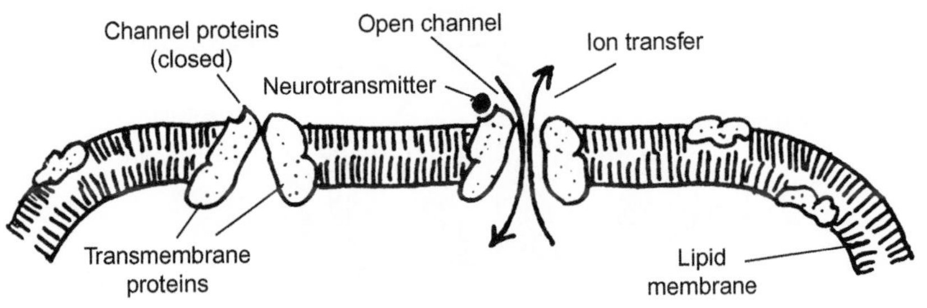

FIGURE 7.3 Pores in the membrane.

substances, such as water and oxygen, can cross the membrane by ***simple diffusion***. However, many of the substances require some sort of transport proteins for their passage across the membrane. The process of rapid diffusion towards the concentration gradient without the expense of energy is called ***facilitated diffusion***. This is carried out by two types of proteins – *carrier proteins* and *channel proteins*. Carrier proteins bind to the molecules and undergo structural changes, allowing molecular transport across the membrane. Examples of carrier proteins are *glucose transporter* and the *erythrocyte anion exchange protein* – but carrier proteins are outside the scope of this chapter.

Channel proteins (Fig. 7.3) are of particular interest in this chapter, as they are the most numerous in neurons. There are chiefly three types – *ion channels, porins* and *aquaporins*. ***Ion channels*** are responsible for the electrical properties of the neuronal membranes, in which they are characteristically abundant. These channels are controlled either by voltage or by ligands (Ch. 2, p. 38). We will study them in the section 'Signal Transduction', below.

The other important transport mechanism is the ***active transport*** (i.e. against equilibrium). This uses energy (in the form of ATP) for their transport; examples are *sodium-potassium pump* (Ch. 2, p. 37), *proton pumps, electron transport chains*, etc.

Finally, very large molecules, such as proteins and fats, can be transported to and fro across the cell by means of ***endocytosis*** or ***exocytosis***, which is not discussed further in this book.

LIPID RAFTS AND MICRODOMAINS

We have seen that the cell membrane consists of two major components – lipids and proteins. Current research shows that the cell membrane is not a monotonous layer of lipids and proteins, but shows localized areas of specialized lipids and proteins. In some places the lipid bilayer shows the

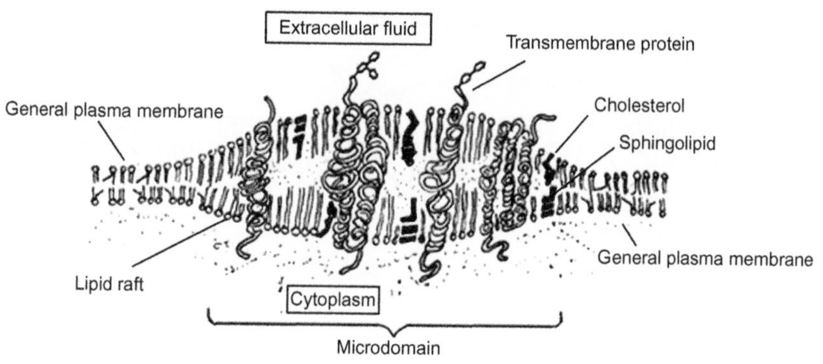

FIGURE 7.4 Membrane microdomain.

presence of a few specialized fats (such as cholesterol and sphingolipids), which make the cell membrane thick at these places – these thickened areas of membrane are called ***lipid rafts*** (Fig. 7.2d). Most often, these lipid rafts host a variety of proteins (especially transmembrane proteins) and these complex units of lipid rafts and proteins are called ***microdomains*** (Fig. 7.4). Thus, a microdomain is a complex mixture of lipids and proteins localized in the membrane.

Microdomains are the structures which give the cell membrane its ***mosaic appearance***. They are known to be actively involved in a variety of cellular functions, especially in cell signaling. It was previously thought that the proteins were the only functional units of a microdomain and that the lipid component was not very important for their function. The outstanding work of Kai Simons and Derek Toomre (2000) has revolutionized the concept of microdomains in cell biology, and showed that the integrity of the lipid component of a microdomain is also essential for membrane function. Cytoskeleton appears to have a role in the organization of microdomains, reflecting its importance in cell membrane function (Hardin et al. 2012, p. 173). The structure of a microdomain is briefly discussed in two parts – the lipid component and the protein component.

The Lipid Component

The ***lipid rafts*** are rich in saturated fatty acids, which are tightly packed together to make a raft thick and rigid. In contrast, the surrounding general plasma membrane is composed mainly of lipids with unsaturated fatty acid tails that are loosely packed (Fig. 7.4) – this makes the surrounding membrane more fluid. This means that the thick sheets of lipid rafts 'float' in the general plasma membrane, like flakes of petals on oil, and are free to move around from place to place on the surface of the cell. This allows a free movement of lipid rafts in the membrane, enabling them to coalesce to form complex groups of microdomains. Perhaps this is the most essential function of the cell membrane.

The size of lipid rafts varies from 10 to 200 nm, and they contain large amounts of cholesterol, sphingolipids and other complex lipids (Fig. 7.4). The lipid rafts may spread flatly on the surface of the membrane (called *flats*), or they may become folded into flask-shaped structures (called *caveolae*). Caveolae are formed by polymerization of *caveolins*, which are hairpin-like palmitoylated integral membrane proteins that tightly bind with cholesterol. Their general function is unknown, but they appear to function as precursors in endocytosis. They are also deemed to assist in signal transduction.

The Protein Component

The number of proteins present in a single microdomain is not very clear, but it appears to be no more than 10–30 (Simons & Toomre, 2000). In general, proteins are not attracted towards lipids, because proteins are polar in nature with carboxyl and amino ends, whereas lipids are non-polar. However, the most important function of a lipid raft is to recruit proteins so that a complex unit of lipid and protein is formed. The proteins can be incorporated into a lipid raft by making them more 'suitable' to stay in a raft. There are two ways by which a lipid raft can accommodate proteins: (1) by changing the lipid composition of a lipid raft because some lipids are more protein-friendly than others; or (2) by changing the protein themselves, by tagging them with some other molecules, such as small lipids or sugars, to form lipoproteins or glycoproteins, respectively. The proteins are bound to the lipid rafts not only by lipid–protein interactions, but also by protein–protein interactions (Fig. 7.6). The importance of proteins in microdomains is further described under 'Genetic basis of primary thoughts', below.

SIGNAL TRANSDUCTION IN NEURONS

Chapter 5 shows that peripheral signals pass up along the nerves in different stages to reach the perceptual neurons in the central nervous system. We will now see exactly where one form of energy is converted into another form before the signals reach the concerned neurons. This process of conversion of energy is called ***signal transduction*** (see Ch. 2, p. 47 & Fig. 2.8). In a sensory pathway signal, transduction can be classified into three levels:

Level 1: the external stimuli may be in the form of light energy (vision), sound energy (hearing), thermal energy (heat), chemical energy (taste, smell) or mechanical energy (touch, pain, pressure, vibration) – all these stimuli are converted into electrical energy at the peripheral receptors and these electrical impulses are transmitted along the nerves. This means that all the external signals are *transduced* into electrical impulses at the peripheral receptors and sent along the nerves. (***Note***: in animals many other stimuli may also be converted by their peripheral receptors into nerve impulses; for example, ultraviolet rays, ultrasound waves, geomagnetic waves, etc., are used by various animals for navigation or predation, as shown in Ch. 4, p. 85)

Level 2: now these electrical signals travel along the sensory pathways. We have seen that there are several synaptic junctions in this journey and at each step the transmission is chiefly *electrochemical* in nature – that is, electrical energy converted into chemical energy and chemical energy converted back into electrical energy (Ch. 2, p. 47 & Fig. 2.8). Here the synapses mediate the transduction of energy at each stage.

Level 3: the signals ultimately reach the perceptual neurons. At the perceptual neurons these electrical signals are finally converted into primary thoughts. In Chapter 6 we saw that it is the dendritic membrane of perceptual neurons that converts signals into primary thoughts. This chapter shows it is the *graded potentials* that are converted into primary thoughts.

Before going into this we will briefly learn about membrane receptors, because it can be said that receptors are crucial for signal transduction.

Receptors in Neurons

Receptors to chemical messengers are present over the surface of *all* cells in the body – both 'ordinary' cells and neurons – except that the distribution of receptors in neurons differs greatly from that of non-neuronal cells. A brief discussion on receptors is presented here without going into details. We will see how the neurons use receptors to convert chemical messengers (neurotransmitters) into electrical signals.

Structure of a Receptor

We have some very detailed knowledge of some kinds of receptors in the cell membrane. Almost all of these receptors are ***transmembrane proteins*** (see above; Fig. 7.3) because they have to receive signals from the external environment and transmit them to the inside of cell. Consequently, each receptor protein consists of three domains: an ***external domain*** – which receives the ligand (chemical messenger); a ***middle domain*** – which is embedded in the fatty layer of cell membrane; and an ***internal domain*** – which 'hangs down' into the cytoplasm and transfers information from outside to inside (Fig. 7.5).

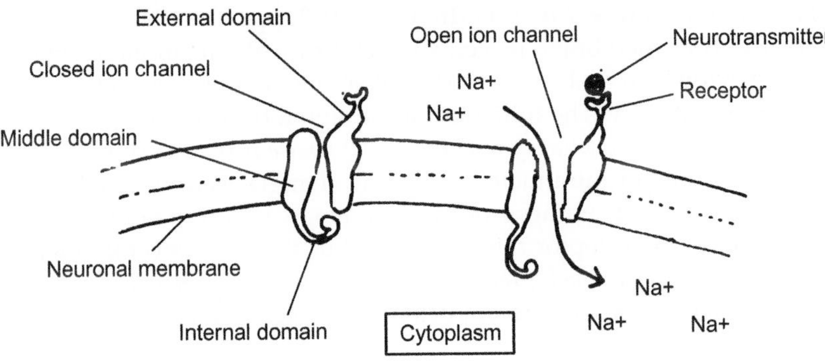

FIGURE 7.5 Ion channel-linked receptor – closed and open states.

Receptors in the Neuron

In the case of neurons the receptors are chiefly concentrated on the postsynaptic ***dendritic membranes*** of the synaptic junctions, and are excited by the neurotransmitters. There are three kinds of receptors present in the neurons:

- ***Ion channel-linked receptors***. The most prominent receptors in the case of neurons are ion channel-linked receptors, which makes them electrically so excitable (Ch. 2, p. 38). They are described separately in the next section, below.
- ***G-protein-linked receptors*** (Hardin et al. 2012, pp. 396–405). These are complex transmembrane proteins that have three subunits (see the reference above for a detailed understanding). GTP activates the α-subunit, which in turn activates the target proteins that initiate a cascade of intracytoplasmic responses, i.e. the signals reach the inside of the cells and cause changes in the cytoplasm and genes. This is mediated by the ***second messenger system***, through *cyclic-AMP* or *$InsP_3$-DAG pathways*, and finally works by activating the *calcium-calmodulin pathway*. *Phosphorylation* and *dephosphorylation* of proteins play a crucial role in orchestrating these events. These changes are ultimately carried into the nucleus where the genes are made to respond and synthesize m-RNA (by *transcription*). The m-RNA crosses the nucleus and enters the cytoplasm where new proteins are synthesized by ribosomes (by *translation*). This new protein mediates a suitable response to the stimulus. The genes can 'remember' these stimulus-response events and retain a 'hard copy' of them, because of which the cell can respond fairly quickly when a signal is repeated. Examples of G-protein receptor-dependent functions are insulin secretion, glycogen metabolism in liver, activation of light-sensitive compounds (in the rods and cones), functioning of various hormones and pheromones, and many more.

- ***Enzyme linked receptors*** (Hardin et al. 2012, pp. 406–409). These are also transmembrane proteins which bring about similar changes in the cytoplasm and genes as described above, but they follow a different pathway. The initiation of the signal transduction cascade here is by the *Ras pathway* and the *MAP kinase pathway*. The MAPK is responsible for intranuclear signaling changes leading to DNA transcription and synthesis of new protein. A prototype of these receptors is the receptor *tyrosine kinase*, but there are several others. Receptors for *epidermal growth factor*, *platelet-derived growth factor*, *nerve growth factor*, etc. are important examples of this type of receptor.

We will concentrate only on the ion channel-linked receptors, as they are concerned with the generation of primary thoughts in the neurons, as we will see in the Mechanism of generation of primary thoughts, below.

Ion Channel-Linked Receptors and Graded Potentials

It is of our interest here to study these receptors in some detail. Neurons possess ion channel-linked receptors in their membranes in abundance, and they are in fact responsible for their electrical activity. Ion channel-linked receptors are ligand-gated channels which are excited by neurotransmitters, and are located chiefly in the dendritic membrane. These receptors are so-named because these channels are associated with ion channels, which open when ligands bind to these receptors (Fig. 7.5). Various ions, such as Na^+, K^+, Cl- and Ca^{2+}, are transmitted across the cell membrane in this way. Common examples are *glutamate receptors* (AMPA, NMDA and kainate receptors, Ch. 3, p. 68), *GABA* receptors, *serotonin* receptors, *glycine* receptors, *acetylcholine* receptors and *P2X*, and their ligands are glutamate, GABA, serotonin, glycine, acetylcholine and ATP, respectively. Upon excitation, these receptors open up their associated ion channels, which cause transfer of ions from one side of the membrane to the other thus creating a concentration gradient. This concentration gradient causes the generation of membrane potentials (MPs; Ch. 2, p. 36).

Chapter 2 takes the example of the excitatory neurotransmitter *glutamate* and studies how ***graded potentials*** are generated at the postsynaptic dendritic membrane. Glutamate binds with AMPA receptors and causes an opening of its associated Na^+ ion channels (see Fig. 6.4). This leads to a rapid inrush of Na^+ ions into the dendrite and the dendritic membrane depolarizes (Ch. 2, p. 37). As the voltage-gated Na^+ channels are not numerous in the dendrites (as in the axons), the shift of Na^+ ions does not result in the generation of action potentials (APs). However, this ionic redistribution does result in producing smaller MPs across the dendritic membrane. The strength of these smaller MPs varies with the number of glutamate molecules bound with AMPA receptors – the greater the number of excited receptors, the higher their voltage. This is the cause of generation of variable MPs called graded potentials (Ch. 2, p. 40).

The Fate of Graded Potentials

Now we will see what happens to the graded potentials – what is their fate? Based on the knowledge we gained from Chapter 5, pp 95–100, we can say that in the case of ***non-perceptual neurons*** the graded potentials generated at the dendrites pass to the soma and join membrane potentials from other dendrites (*summation*) and collectively act on the axon hillock (Fig. 2.5). The axon hillock then fires APs and the signals are passed on to the next neuron, and the chain continues until they reach the perceptual neurons.

We will now see what happens to the graded potentials in the case of a ***perceptual neuron***. These graded potentials must now generate primary thoughts in the perceptual neurons by interacting with the molecular grids situated in the dendritic membrane. Before going into this we will see what these molecular grids are, and what their structure is.

ANATOMY OF MOLECULAR GRID

We have seen that certain molecular gadgets are necessary to convert external stimuli into primary thoughts, and we have also seen that they are located in the dendritic membranes of perceptual neurons. We will now look into their structure.

It is shown in above that the cell membrane consists of localized thickenings called ***membrane microdomains***, which are made up of stacks of specialized lipids (lipid rafts) along with transmembrane proteins. Basing on these findings we propose a new concept, the ***molecular-grid model***, for the generation of primary thoughts by the neurons.

What is a Molecular Grid?

A molecular grid is formed by a group of microdomains forming a complex unit, which is responsible for the conversion of external signals that reach them in the form of graded potentials into primary thoughts.

How are these microdomains that constitute a molecular grid held together? The proteins in the microdomains carry a net electrical charge over them owing to the presence of polar amino acids. These charges facilitate the formation of various molecular interactions between the proteins, which vary in their binding strength depending on the type of interaction

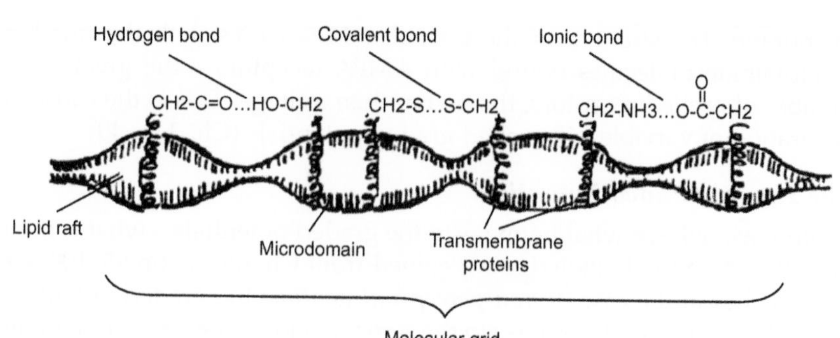

FIGURE 7.6 Molecular-grid showing protein–protein interactions.

they form (Fig. 7.6). The various lipids also interact with proteins in the form of various chemical bonds to keep these molecules together. The most common types of molecular interactions between proteins are described as follows (Hardin et al. 2012, pp. 44–47):

1. ***Covalent bonds***: most commonly these are the *disulfide bonds*. They form very strong bonds (70–100 kcal/mol), hence they are very stable for holding proteins together. They presumably hold proteins together for a long time.
2. ***Ionic bonds***: these are electrostatic interactions between positively charged and negatively charged amino acids. They are not as strong (about 3 kcal/mol) as covalent bonds.
3. ***Hydrogen bonds***: these are interactions between a hydrogen atom and a pair of electrons of an oxygen atom. Their bond energy is about 2–5 kcal/mol. These bonds can possibly hold proteins for variable periods of time.
4. ***Van der Waals' interactions***: these are very weak and transient bonds arising from asymmetric distribution of electrons in various atoms (0.1–0.2 kcal/mol).
5. ***Hydrophobic interactions***: they are not real bonds but are indirect forces between non-polar molecules in an aqueous medium. The non-poplar entities are generally fatty acids or non-polar amino acids.

All these interactions 'chain down' the various proteins in a microdomain with variable strengths. The significance of this becomes apparent once we study the molecular basis of memory in Chapter 8 (p. 153). Obviously, the fluidity of the cell membrane is the most important feature in understanding the kinetics of a molecular grid – a rigid membrane can neither bring microdomains together nor cause them to drift away from each other – hence the neuronal membrane must be maintained in a fluid-like phase by the unsaturated fatty acids (see Structure of cell membrane, above).

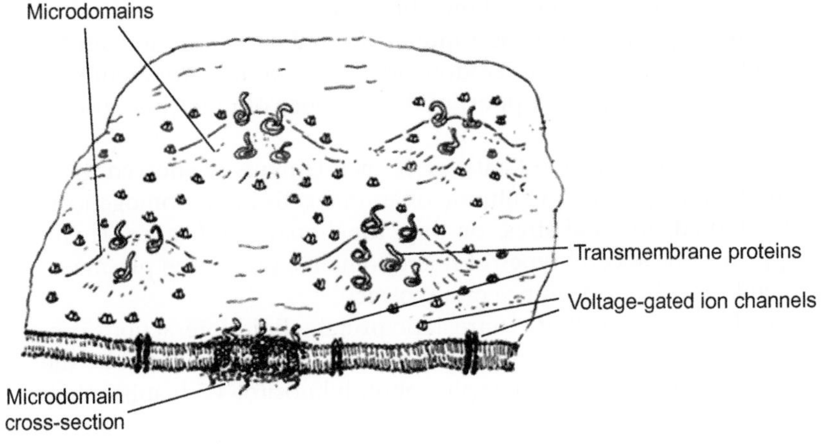

FIGURE 7.7 The molecular grid.

It has been documented that *APs* are recorded in dendrites (Barrett et al. 2012, p. 125), which stands as evidence for the existence of *voltage-gated ion channels* in the dendrites. It appears that molecular grids are closely associated with the voltage-gated ion channels (Fig. 7.7), and the importance of their presence becomes obvious in the section on 'thought-potentials', below.

Proofs of Molecular Grids

At one time in the past the existence of microdomains was debated, but now their existence is well established – and it is shown that they have an active participation in signal transduction. The following research studies are cited as three examples (among many others) that support the existence of molecular grids in the neuronal membranes:

1. Scientists at Max Planck Institute of Biochemistry in Munich (Spira et al. 2012) have performed a comprehensive analysis of the entire cell membrane and its independent domains. They found that if a protein is relocated to an inappropriate domain, the membrane may fail to process signals and thus does not respond to changing conditions (see *'barcode arrangement!'* below). They have proved that microdomains are organized into *'bounded domains'* and state that domain formation is not an exception but is the rule. Each protein is located in distinct areas that are patterned like a ***'patchwork quilt'***, with some areas containing more than one type of protein. The big question is: how do the proteins find an appropriate domain for them? The lipids in the cell membrane appear to play a central role – different lipids prefer to accumulate around certain protein anchors, and this explains how microdomains self-organize. Apparently, these findings stand out in support of the presence of molecular grids in the membranes.

2. Professor F.R. Maxfield (2002) of Cornell University showed that the lipids in the plasma membrane of animal cells are inhomogeneously distributed, and laid stress on their functional significance. There appear to be several types of microdomains, giving the plasma membrane the appearance of a mosaic.
3. It is demonstrated that electrostatic protein–lipid interactions can result in the formation of microdomains, and are mediated by the strongly anionic lipid PIP2 (phosphotidylinositol-4,5-bisphosphate) (van den Bogaart G et al. 2011).

MECHANISM OF GENERATION OF PRIMARY THOUGHTS

In the above sections we have seen the mechanism of generation of graded potentials and studied the structure of molecular grids. Now we will see how the graded potentials interact with the molecular grids in order to produce primary thoughts. We will take the example of the thought-analysis described in Chapter 4 (p. 78) and explain how the primary thought of an 'unnamed color blue' is generated in the dendritic membranes of a perceptual neuron in the visual cortex.

When blue light impinges on the retina, S-cones are excited, which send signals to the ganglion cells; these ganglion cells fire APs and send them to the neurons of the lateral geniculate body (see Ch. 3, p. 55 & Fig. 3.2). The nerve impulses from here reach the perceptual neurons in the visual cortex in the form of APs. These APs reach the synaptic junctions over the dendrites of the perceptual neurons and cause a release of NTs into the synaptic cleft, which act on the concerned receptors in the dendritic membrane (see 'Signal transduction in neurons', below). The ion channel-linked receptors in the dendritic membrane produce an ionic flux, thus generating small graded potentials.

The perceptual neurons concerned here house the molecular grids necessary to generate the primary thoughts of blue color. The graded potentials traverse along the dendritic membrane and interact with the molecular grids, which when excited generate primary thoughts of blue color. We will now see a mechanism by which this interaction takes place.

The 'Thought-Potentials'

To understand how the molecular grids are excited by the graded potentials, we must first study the electrical properties of proteins, because proteins are the essential components of a molecular grid. All proteins carry varying charges on them, depending on the amino acids they contain. Some amino acids are acidic in nature, some are basic, some are polar and some are non-polar, thus the net charge a protein carries depends on its constituent amino acids. Also, special proteins such as acylated or prenylated proteins are aplenty in the cell membranes, which may differ in their electrical conductivity. Thus the overall electrical property of a molecular grid itself is dependent on its protein composition. Consequently, it can be said that the protein-rich microdomains in a molecular grid are relatively good conductors of electricity compared to their adjacent lipid rafts, which are obviously poor conductors of electricity. Thus, in effect, we have two regions in a molecular grid with two distinct electrical properties:

- Protein-rich microdomains, which are relatively good conductors of electricity
- Lipid-rich lipid rafts, which are poor conductors of electricity.

We can now see that there are good conductor regions alternating with poor conductor regions in each molecular grid. It can be presumed that each molecular grid has a characteristic arrangement of microdomains and rafts; this arrangement is peculiar to each of them and makes each molecular grid distinctly individual – similar to the barcode we see on the retail products and ID tags (Fig. 7.8a).

Thus, when graded potentials traverse a molecular grid, it can be presumed that each molecular grid responds with a specific electrical

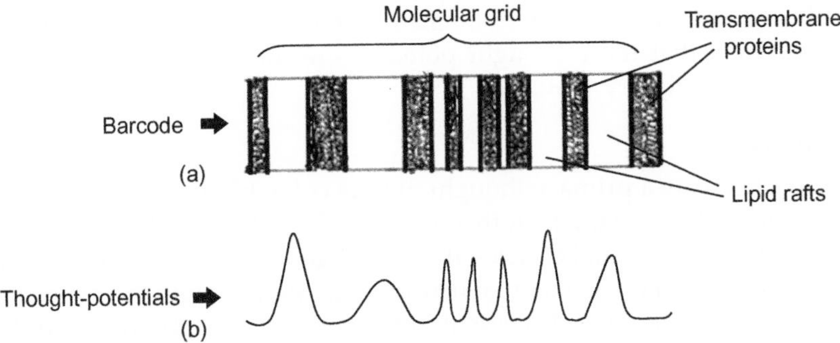

FIGURE 7.8 (a) Dendritic membrane with a 'barcode' arrangement; (b) with corresponding thought potentials.

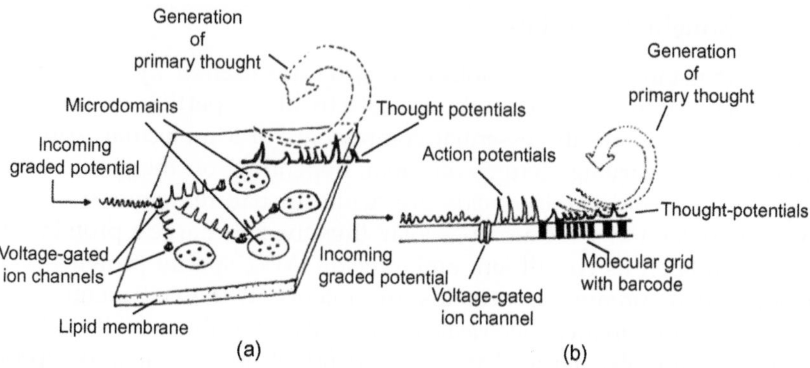

FIGURE 7.9 Schematic diagrams showing generation of primary thoughts.

pattern, depending on the characteristic arrangement of these alternating microdomains and lipid rafts. The electrical signal that passes through a protein-rich microdomain gives a 'flat trace', whereas the electrical signal that passes a poorly conducting lipid raft 'jumps' (saltatory conduction), i.e., it gives out a *wave form* (Fig. 7.8b). These potentials are called ***thought-potentials***. The pattern of these thought-potentials obviously depends on the pattern of the barcode, and obviously each molecular grid gives out a different wave pattern depending on its protein composition. Now we can say that these molecular grids have acquired specificity owing to the presence of the barcodes.

We have shown, above, that there is evidence of the presence of voltage-gated ion channels in the dendritic membrane. It can be presumed that each microdomain in a molecular grid has a few associated voltage-gated ion channels surrounding it (Figs 7.7 & 7.9). Perhaps the function of these voltage-gated channels is to produce action potentials of adequate amplitude to excite all microdomains in a molecular grid. Finally, it can be presumed that these thought potentials result in the formation of a characteristic primary thought.

Summary

Thus, to create a primary thought of blue color, blue light has to excite an S-cone in the retina, which then sends a signal to the relevant perceptual neuron in the visual cortex – this signal generates a graded potential in the dendrites of the perceptual neuron that interacts with the molecular grid with a specific barcode present in the dendritic membrane – and this gives out a thought-potential, which again is converted into a primary thought of blue color. This is the overall mechanism by which a ***sensation*** of blue at the receptors is converted into a ***thought*** of blue at the neurons. All the sensations, whether general or special, are converted

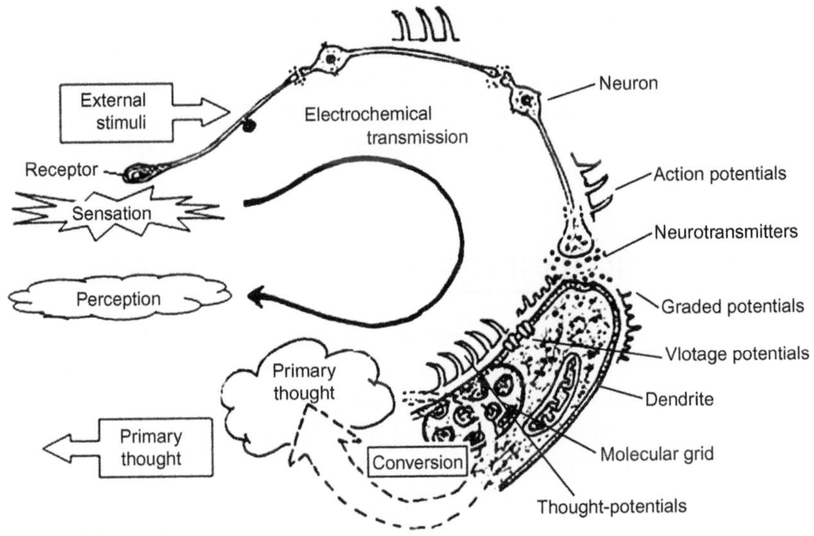

FIGURE 7.10 Sensations to perceptions – an overview.

into thoughts by similar steps, as summarized in Figure 7.10. Obviously, each of these senses is represented by a specific molecular grid that generates a specific thought-potential, and this specific thought-potential is translated into a specific primary thought (see 'Genetic basis of primary thoughts', below).

What is a Primary Thought?

What is the nature of thought? Is thought some kind of energy – if so, what form? We will now see the various forms of energy that the neuron is capable of dealing with. The energy transmitted along the nerves and neurons is electrochemical energy generated by ions, as shown in the preceding chapters, and the graded potentials in the dendrites also represent this form of energy. The brain utilizes a lot of glucose and oxygen to generate electrochemical energy to maintain polarity across the neuronal membranes (see Ch. 2, p. 52). We have seen that this energy finally interacts with the molecular grids to generate thought-potentials. If you think of the neuron as a thermodynamic system, the energy that is supplied to its billions of neurons must ultimately be dissipated into its surroundings. But in what form does this energy leave the neurons? Presumably, the majority of this energy is converted into primary thoughts (after all, the function of the neurons is the generation of thoughts), and some energy is invariably lost into the surroundings as heat (*entropy*) – as in the case of light and heat in an incandescent bulb.

Whereas graded potentials represent ionic energy with an estimated conduction velocity of a modest 4–120m/sec (Ch, 2, p. 40), primary thoughts may represent some form of hitherto unidentified electromagnetic energy (with a speed of *299 792 458 m/sec*!). However, the nature of thought is a complex metaphysical issue, best reserved for later discussion in Chapter 10.

GENETIC BASIS OF PRIMARY THOUGHTS

As we can see, there are several perceptual neurons in different areas of the brain, which can generate different types of primary thoughts; for example, visual neurons can produce perceptions of color, olfactory neurons the perceptions of smell, and on. This means that each of these neurons houses a different set of molecular grids to generate different primary thoughts.

How do the molecular grids generate different primary thoughts? As was shown above, the answer lies obviously in the proteins that make up a molecular grid. The molecular grid has a specific 'barcode' arrangement of microdomains and lipid rafts that gives a characteristic pattern of thought-potentials. These differences in their electrical properties make each molecular grid distinctly individual – a specific barcode for each primary thought.

Quite obviously the type of proteins present in a molecular grid is determined by the genetic make-up of the individual. Thus, we can arrive at the conclusion that it is ultimately the genes that generate different types of primary thoughts.

By way of giving an example, we can presume that to taste the sensation of *salt*, we must possess a different barcode in the molecular grids lodged in the dendritic membranes of taste-sensing perceptual neurons in the insular lobe of the brain – these generate different primary thoughts related to 'salt' sensation. And those neurons that are designed to decipher the sensation of *sweet* have a different barcode in their molecular grids producing different primary thoughts of 'sweet' sensation. Likewise, each sensory modality (color, smell, sound, pain, touch etc.) has a different barcode to generate an appropriate primary thought.

In Chapter 4 we saw that primary thoughts are inherited and that genes are responsible for the generation of various kinds of primary thoughts. In the above sections, we have discovered the physical basis for the inheritance of primary thoughts in the form of proteins and genes. The genome in a neuron (the whole set of genetic information in a cell) synthesizes all the proteins needed for the cell, and thus the genome is responsible for the various barcodes in a molecular grid. It is natural to suppose that we have a 'bank' of molecular gadgets in the neurons, at our disposal at the time of birth – they are already lodged in the basic structure of an individual's mind. This explains the phenomenon of ***a priori*** described in Chapter 4, which shows that primary thoughts are built-up in the fundamental structure of the mind and need no prior experience for sensing them. This is the reason that we are able to perceive blue color or red color, hear sounds of various frequencies, feel pain or pressure, taste salt or sugar, and so on. And this is also the reason that we *cannot* perceive UV rays – the molecular grids that can generate primary thoughts of UV light are simply absent in human neurons (owing to the fact that humans lack the genes that can synthesize the specific protein needed for these barcodes, but presumably some animals can synthesize these characteristic proteins). This is the genetic basis of primary thoughts and ultimately that of our mind and intelligence, as we shall see in Chapter 8, under Intelligence.

MOLECULAR GRIDS AS 'BIOLOGICAL TRANSISTORS'

Molecular grids can be compared to electronic circuits regarding their electrical properties such as resistance, conductance and capacitance. The electrical potentials that propagate in the dendritic membranes can be modulated in several ways by the molecular grids, and can be utilized for the generation of primary thoughts. It is interesting to see, in this context, a close relationship between the structure of a molecular grid and a transistor in a modern computer.

A transistor has three layers of semiconductors – an emitter, a collector and a base, and it controls the electric conductivity across it to perform a desired task (Ch. 11, p. 208). A molecular grid also can be considered to have three layers of conducting media with variable electrical properties – the ECF, the cytoplasm and the membrane itself (Ch. 2, p. 36) (Fig. 7.11), and here also the incoming voltage potential is 'modulated' to generate various kinds of primary thoughts. Hence, a molecular grid

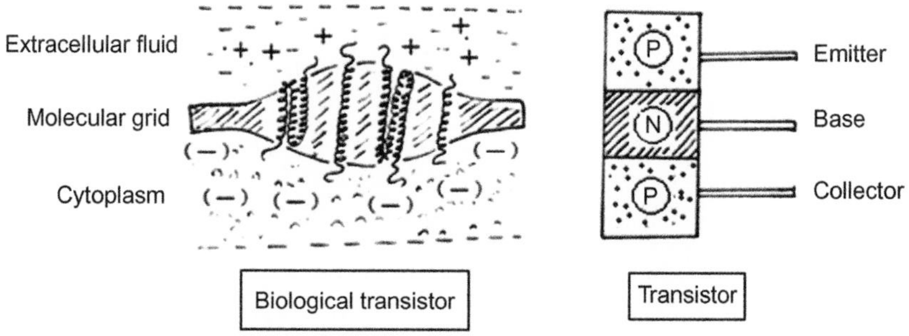

FIGURE 7.11 Comparison between molecular grid and transistor.

may be called a *'biological transistor'*. The reader can find an interesting discussion of this comparison in Chapter 11.

Now, at last, we can see that the hitherto abstract thought has a physical basis in the neurons and can in fact be studied not only by biologists, but also by physicists, who can include the study of thought in the quantum realm. Thought has finally become an interdisciplinary science!

In Chapter 8, we will look at the possible implications of this model. As with any good working model, the molecular grid model should explain, at least in part, the major phenomena of the brain. We saw in Chapter 3 that the most fundamental phenomenon of the brain is memory, and thus we will apply the molecular grid model to explain the molecular basis of memory and other related phenomena.

CHAPTER

8

Memory, Intelligence and Molecular Grid

OVERVIEW

Chapter 7 describes the molecular-grid model by which external stimuli are converted into primary thoughts. To test the validity of this model, this chapter will try to explain the molecular basis of the fundamental phenomenon of the brain, i.e. memory, by using the molecular-grid model. We studied memory traces in Chapter 3, but we have not known the exact mechanism by which molecular signatures are formed. In this chapter we will see where exactly a group of neurons store memory and which mechanisms are involved.

Based on this mechanism, this chapter will also examine the mechanism of memory retrieval and forgetfulness, and try to explain a few other complex phenomena of the brain, such as intelligence and some other mental eccentricities, for example *déjà vu* and child prodigies. This model also explains how genes determine the intelligence of an individual, and why certain special fats such as docosahexaenoic acid (DHA) improve intelligence. It becomes obvious, as we read along, how well this model explains all these phenomena aptly and precisely.

First, we will outline the progression of memory – how the sensory memory progresses to short-term memory (STM) and then to long-term memory (LTM). Then we will see how memory traces are formed, stored and retrieved. The later sections of this chapter describe phenomena such as forgetfulness, intelligence and other such related mental phenomena.

MEMORY: A REVIEW

As described in Chapter 3, in 'Formation of memory' (p. 57), human memory is chiefly divided into four categories – sensory memory, STM, working memory and LTM. Let us study memory afresh in the light of

The Biology of Thought.
DOI: http://dx.doi.org/10.1016/B978-0-12-800900-0.00008-7

the new molecular-grid model. Throughout this chapter the reader may refer to this part of Chapter 3.

Progression of Memory

Imagine an artist looking at some beautiful scenery, with a wonderful blue mountains, a lake below, a speeding motor boat in the center of the lake, flowery bushes to one side and a dinghy lying motionless in one corner of the lake. The artist is painting this landscape on a canvas. Now, you may ask him: 'What are all the objects you have observed in the scenery?' We will take the experience of our artist and try to understand all types of memory. The progression of memory may be described in four steps:

Step 1. When our artist looks at the scenery the images of all the details impinge on the retina and are converted into ***sensations***. These sensations are transmitted to the brain where they are converted into ***perceptions*** (see Fig. 3.1). When asked the above question, the artist may describe all the details of the scenery but may fail to mention the presence of the dinghy, because it is lying in one nondescript corner in the lake so has escaped his notice. Surely, the image of the dinghy also has fallen on to the retina along with the other details and has been transmitted to the brain? But our artist has failed to register the dinghy in his memory. All these images that have impinged on the retina are transmitted to the brain, irrespective of their being registered in the conscious mind or not. This is the ***sensory memory***. Sensory memory is nothing but an array of perceptions as they are received from the external world. Sensory memory is the *first step* in the formation of any memory – and it is nothing but perception itself. It is claimed that this memory is retained for about a few hundred milliseconds to 1–2 seconds, before being either converted into STM or vanishing from the mind (as in the case of the dinghy, above; see Fig. 3.4).

Step 2. However, the artist has taken notice of the mountains, lake, flowers, speeding boat, etc., hence these features are stored in the brain for a transient period. This is called ***STM***. In other words, those details which have entered the conscious appreciation are said to be incorporated in the STM. The duration of its storage ranges between 10 and 12 seconds. It is obvious that the fate of this memory depends on the attention the perceiver pays to it – if no attention is paid, the memory decays (Fig. 3.4); if conscious effort is made, it forms LTM. STM appears to be the property of all animals with a nervous system; even a snail has to execute some sort of STM in order to move around in space (Ch. 9, p. 167).

Step 3. Our artist is painting a picture on canvas. He first looks at the scenery, stores it in STM, looks back at the canvas and uses the brush to paint the picture – he does this gradually by intermittent observation and repeated brush strokes. This sequential *multitasking* needs temporary storage for a slightly longer duration than mere STM and in a better organized fashion for efficient execution. This is called ***working memory*** (p. 58). It can be shown that working memory also includes some LTM, because the act of executing of even a simple task encompasses several experiences of the past, some long-held beliefs and prejudices, too. For example, in order to depict a speeding boat, our artist has to paint trailing watermarks behind the boat – the knowledge to paint in this way is a part of LTM; hence, working memory is as complex as LTM. Working memory appears to be the prerogative of only human beings and perhaps some higher animals.

Step 4. After finishing the painting, the artist may be commissioned for some other task. But the he may remember the scenery and this painting for a variable period of time, depending on the interest he takes in it. Obviously, this is ***LTM***. Its duration may be a few hours, days, years or lifelong.

Now we will analyze the molecular basis of STM and LTM by taking into account our knowledge of perceptual neurons, dendrites and molecular grids.

MOLECULAR MECHANISM OF MEMORY

Chapter 3 outlined the importance of various areas in the brain, such as several cortical areas, limbic system, hippocampus, Papez circuit, etc., in the formation of LTM. Now, in the following sections, we will see precisely where the primary thoughts are stored in the neurons and which mechanisms are involved.

Formation of Memory Traces

It was ascertained in the preceding chapters that primary thoughts are generated in the dendritic membranes of the perceptual neurons. We can now say that to form a meaningful idea of an external object, several primary thoughts have to be arranged in order. For example, if you have to *see* a flower, each bit of color from the flower, each shade of color, has to be sensed by the retina and sent to the brain for conversion into primary thoughts, and all the primary thoughts of shades thus reached the visual cortex have to be assembled in an order to form an idea of a flower. This array of primary thoughts jointly gives rise to an idea, which is stored as memory in the brain. This STM is somehow *structurally depicted* in the neurons of cortex; this structural unit of memory is called a ***memory trace*** (see example, below).

Memory trace is the hardware – STM is the software

Hitherto it was *presumed* that these memory traces were represented by a network of synapses (Guyton 2008, p. 725; Barrett et al. 2012, p. 286) and that the synapses were the repositories of memory. Now, based upon what we have seen in the previous chapters we will examine STM and LTM in a new light. We will take the following simple event for our analysis.

It is a common experience that memory traces are formed very quickly, and once formed can give you an *instantaneous recall*. As an example, if you are shown an oddly shaped chip for a short period of time and then mix that chip in a pile of 20 different chips, you can pick the odd chip out instantly (explanation below). The memory of that odd chip was 'etched' on your mind, and formed a memory trace (and resulted in the formation of STM). This memory trace has enabled you to pick it up again from the pile *instantaneously*. Below we will see the structural basis of this memory trace. (Of course if you *want* to remember this odd chip for a longer time, the memory trace has to be stored in LTM.)

Many Neurons are Needed

Human thinking generally comprises several types of primary thoughts forming various complex ideas (Ch. 4, p. 83) – a casual object observed, like a red ball on the ground or the pen you are holding, is the result of a mixture of several different primary thoughts. We have seen that ***no single neuron*** can hold all the thoughts needed to build up even a simple image of the external world (Bruno 2002; and see Ch. 3, p. 66). That is to say that to sense even a simple object or sensation (e.g., a flower, the flavor of ice-cream, the sensation of pain or a piece of music, etc.), a single primary thought generated by a single neuron is grossly insufficient. Moreover, all of our experiences are composed of not one type of primary thoughts, but a variety of them – thus, even simple experiences need different neurons to work jointly. Only rarely does our thinking consist of only a single type of primary thought, as in the case of monochromatic blue light (Ch. 4, p. 78) or perhaps the primary thoughts of a single modality of smell or taste where they comprise a single type of primary thought. But even these initial primary thoughts are immediately converted into simple or complex ideas ('a blue sky', 'a rose-like scent', 'a sugary taste', etc., as seen in Ch. 4, p. 79).

All Neurons are Connected

Thus, human memory is always the result of the work of several hundred or perhaps even thousands of *cross-talking* neurons. In other words, for the memory traces (= STM) to form, a group of neurons is needed. All the neurons have to cross-talk with each other ***simultaneously*** in order to produce a memory trace. In 'Storage of memory and recall', below, we will see how this happens.

The Problem of Synaptic Delay

The above example shows that it has taken a very short time to store the memory trace of the odd chip, and an even shorter time to recollect the memory of it. A fleet of neurons have worked together ***collectively, simultaneously*** and ***instantaneously*** to give the desired result. Each primary thought (or percept) was 'knitted' to the other primary thought without significant delay in order to form a ***contiguous*** or ***holistic*** image of an object. This has given you a 'flash memory' of the odd chip.

In the preceding chapters we described that all this transfer of information between the neurons is carried out chiefly by ***axodendritic*** synapses, putting their chemical synapses at work. We know that there is some significant ***synaptic delay*** (see 'Chemical synapses', below) in transmission of impulses in the case of chemical synapses from one neuron to the other. It has been shown that the greater the number of synapses in a neural circuit, the greater is the synaptic delay (Guyton 2008, p. 570). But then, the formation and recollection of the odd chip was *almost instantaneous*. There was apparently no such delay in the re-formation and recollection of memory trace. What happened to the synaptic delay? How have the neurons surmounted this delay?

Synaptic delay can be surmounted in two ways:

- By employing ***dendrodendritic synapses***, *bypassing* axodendritic synapses (see below)
- By using ***electrical synapses***, *bypassing* chemical synapses (see 'The role of electrical synapses', below).

Now we will see how both of these work out in shortening the synaptic delay.

Dendrodendritic Synapses

We have seen several types of synapses in Chapter 2. We have also seen that axodendritic synapses are the most common type of synapses in the central nervous system (CNS; Ch. 6, p. 110). Ordinarily a signal from the dendrite has to pass to its soma and thence to the axon, from where it ends on the dendrite of the next neuron – this traditional view of ***unidirectional flow of information*** is discussed in Chapters 2 and 6. However, it can be said that by bypassing this 'loop' a signal can spread fast from one neuron to the other.

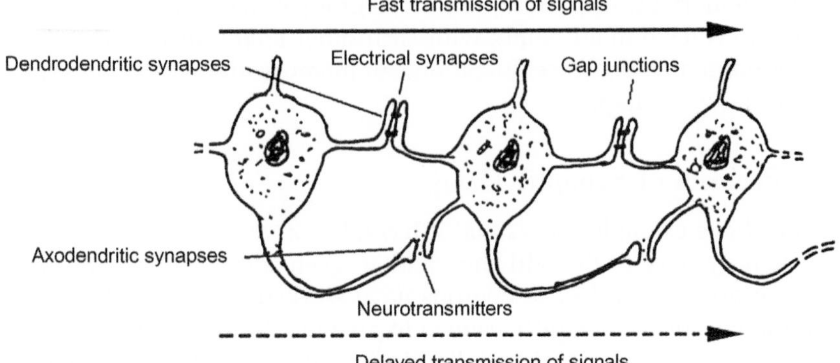

FIGURE 8.1 Axodendritic synapses *versus* dendrodendritic synapses.

One way to bypass this loop is to employ ***dendrodendritic synapses.*** This synapse directly connects dendrites of two neurons sending signals from one dendrite directly to another dendrite, bypassing the axodendritic synapse (Fig. 8.1) (Fine 2008, p. 29). This sort of efficient transfer of information is required in a place where there is a premium on space and time, as in the case of memory storage and recall in the cortex.

It is no wonder that current research has identified dendrodendritic synapses in virtually all places in the CNS – cortex, diencephalon, brainstem, cerebellum, spinal cord. *Presumably, all the perceptual neurons located in the CNS are connected this way.* These synapses are also identified in places like the retina, where *bipolar cells* communicate in this way with *amacrine cells* (Ch. 3, p. 54).

It has been thought that dendrodendritic synapses are also chemically mediated by excitatory and inhibitory molecules (much like the axodendritic synapses). But, it appears that dendrodendritic communication is *not* chemically mediated, because some very fast synaptic communication is needed for efficient transfer of information from dendrite to dendrite. It is possible that they communicate more efficiently by means of ***electrical synapses*** as discussed below.

The Role of Electrical Synapses

Theoretically speaking, there are three possible ways by which neurons communicate with each other:

- Chemical synapses (through synaptic junctions)
- Electrical synapses (through 'gap junctions', see below)
- Ephaptic coupling (without any synapses).

TABLE 8.1 Differences Between Chemical and Electrical Synapses

S. No	Chemical Synapses	Electrical Synapses
1. Functioning	Complex	Simple
2. Synaptic cleft width	30–40 nm	3.5 nm
3. Junctions	Synapses	Gap junctions
4. Speed of transmission	Slower (relatively)	Faster (instantaneous)
5. Neurotransmitters and receptors	Yes	No
6. Synaptic delay	>0.5–4 ms	<0.2 ms
7. Directionality of transmission	Unidirectional	Bidirectional
8. Postsynaptic modulation	Possible	Not possible
9. 'Gating' the signal	Yes	No
10. Plasticity	Essential feature	?
11. Common sites	Chiefly CNS neuromuscular junction, etc.	Heart muscle, smooth muscle, CNS, retina etc.

Chemical Synapses

Chemical synapses use *neurotransmitters* for transmission. The synapses described in the previous chapters are chemical synapses. Though relatively slow, chemical signaling is essential in the CNS for four reasons (see also Table 8.1):

1. They act as gates → to allow or check signals to pass
2. They are unidirectional in passing signals → signals reaches specific targets
3. They are ***not*** all-or-none and so their magnitude may vary according to the amount of neurotransmitter released → the signal strength may be titrated
4. The signals may actually 'gain' strength as they pass across the synapse by increasing the amount of neurotransmitter released → the signal is *augmented* as it reaches the target.

However, the problem with these chemical synapses is the ***synaptic delay***. It is calculated that a certain amount of time elapses for each signal to reach from one neuron to the other through their synapses. This is called synaptic delay. The *minimum* delay that can be expected is *0.5 milliseconds* (Guyton 2008, p. 570), or it can be even up to *2–4 ms*. In fact,

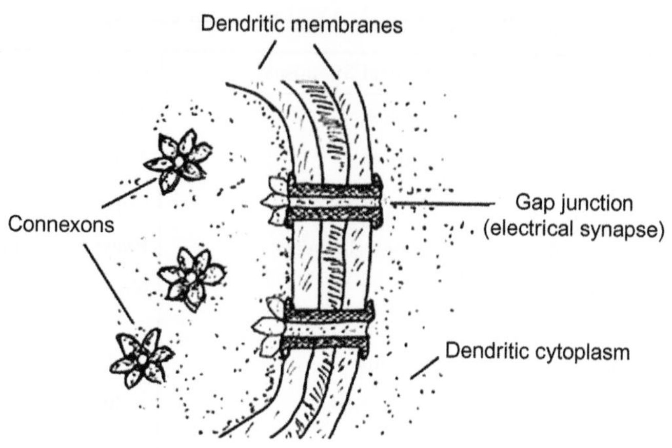

FIGURE 8.2 Electrical synapse.

the number of synapses present in a polysynaptic neural circuit can be calculated by considering the total synaptic delay that has occurred – more synapses, more delay.

Electrical Synapses

An electrical junction allows direct ionic (or electrical) contact between neurons. The cell membranes of the two neurons are connected by specialized junctional proteins called ***gap junctions*** (Fig. 8.2). These gap junctions are made up of tubular proteins called *connexons*, which are in turn made up of subunits called *connexins*. The electrical synapses are designed to conduct impulses at a very rapid rate of less than ***0.2 ms*** (or almost instantaneously), by means of rapid ionic transfer. This means that the electrical connections at the dendrodendritic synapses allow a fleet of neurons to work ***collectively***, ***simultaneously*** and ***instantaneously***, as is required to form an STM or a memory trace.

The differences between chemical and electrical synapses are highlighted in Table 8.1.

Ephaptic Coupling

Apart from chemical and electrical synaptic communication, there exists a third type of communication between the neurons, called ephaptic coupling. Here the communication is 'extracellular' in the sense that there is ***no*** direct contact between the neurons, and by this mechanism it is possible that a neuron may influence other neurons from a long distance (*'remote control'*). Endogenous brain activity is known to be affected by extracellular electrical fields, which may induce membrane potentials without signaling from the neuron above (Anastassiou et al. 2011). The significance of ephaptic coupling is not very clear at present.

Dendritic Electrical Signaling

From the above discussion it is evident that the transmission of information in the neurons of the CNS without synaptic delay is possible by *dendrodendritic electrical signaling* (or simply called ***dendritic electrical signaling***). Nowadays, it is actually speculated that electrical synapses appear to be employed in large networks of neurons to act in a more synchronous manner (for example, similar types of neurons in the cortex are shown to be electrically connected to their identical neighbors). We can presume that all the ***perceptual neurons*** in the layers of ***neocortex*** are connected this way. Now we will see how dendritic electrical signaling is utilized in the storage and retrieval of memory.

STORAGE OF MEMORY AND RECALL

We have seen that primary thoughts are generated by molecular grids located in the dendrites and that several primary thoughts are needed to form simple ideas (see Ch. 4). In other words, the molecular grids must be arranged in a sequential fashion so that they can be excited ***collectively, simultaneously*** and ***instantaneously*** to generate a series of primary thoughts, and this gives rise to a simple idea. How are the molecular grids arranged in series?

Now we will see one such mechanism by which molecular grids are arranged in such a fashion to generate an orderly sequence of primary thoughts.

Dendritic Pleats

For the primary thoughts to be generated serially to give rise to simple ideas, the dendrites containing the necessary molecular grids must be arranged in *arrays*. This sort of arrangement is possible if various neurons are connected by their dendrites in something like 'matted' sheets spread across the cortex – in fact, the dendrites are known to be arranged in densely packed units with similar orientation in the layers of cortex (Barrett et al. 2012, p. 272). This arrangement may give the dendrites a plaited appearance, hence they are called ***dendritic pleats*** (Figs 8.3 & 8.4). These dendritic pleats represent ***memory traces*** in *physical* form. Thus we can modify the statement given on p. 148, into the following:

Dendritic pleats are the memory traces – STM is the software.

Thus, it is implied that a memory trace can store a memory for longer periods if the dendritic pleats stay together for longer periods; conversely, a memory trace can be short-lived if the dendritic pleats are transiently

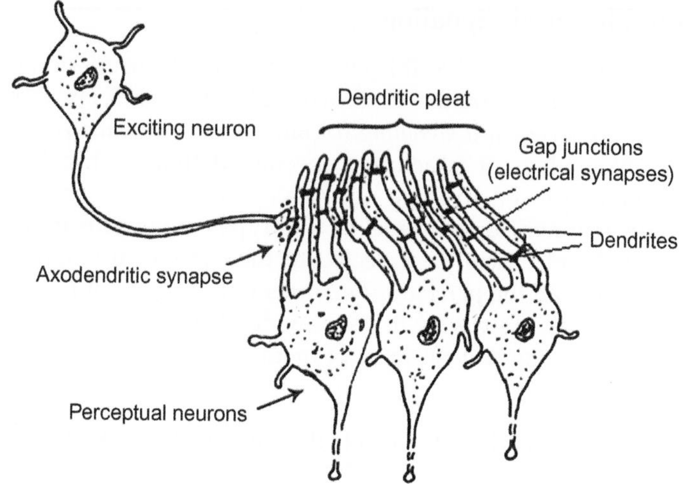

FIGURE 8.3 A dendritic pleat.

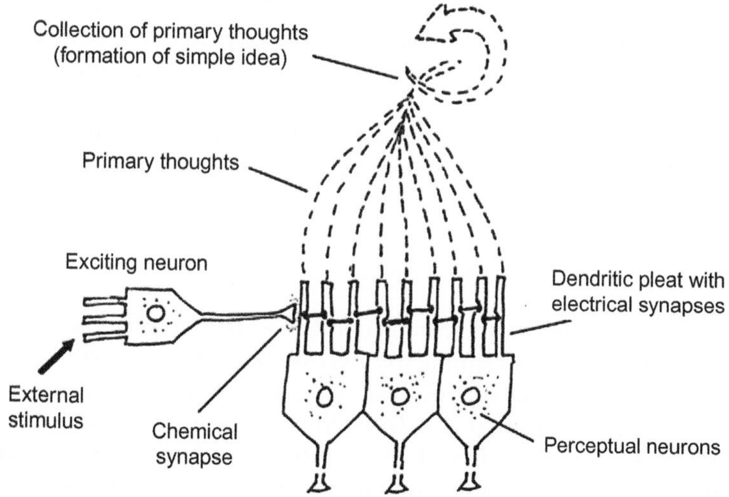

FIGURE 8.4 Dendritic pleat: a schematic diagram.

formed – longer for LTM, shorter for STM. The memory trace can thus be stored for variable periods of time by making the dendritic pleats either transient or long-lasting. How the dendrites are bound down and which forces are involved is discussed in 'Mechanism of durability of memory', later in this chapter.

Formation of Dendritic Pleats

Each sensation we receive from the external world initiates the formation of a dendritic pleat. Incoming stimuli excite a set of molecular grids in the membrane of dendrites to generate primary thoughts. Thus, we can presume that an incoming sensation consisting of various stimuli excite the several molecular grids at the same time. This simultaneous excitation can bring all the concerned molecular grids together, and this results in the formation of dendritic pleats (Fig. 8.3). It can be stated that we have here a variant of *Hebbian process* (Ch. 3, p. 69):

> Molecular grids that fire together, wire together, forming dendritic pleats.

However, for all the dendrites in a dendritic pleat to be excited simultaneously, they must be connected by electrical synapses, so that all the primary thoughts are generated almost instantaneously (Figs 8.3 & 8.4). This explains how an object we sense from the external world generates a series of primary thoughts resulting in the formation of an idea.

Activation of Dendritic Pleat

How do we recollect memories? A dendritic pleat once formed needs an initiating stimulus for it to get excited and generate the concerned primary thoughts and form an idea. We have shown in Chapter 6 that a sensation from the external world reaches the CNS by the way of axodendritic synapses. It appears that this incoming stimulus (transmitted by a chemical synapse, i.e. axodendritic synapse) first excites one dendrite in a dendritic pleat (Figs 8.3 & 8.4) and then the signal spreads throughout the dendritic pleat *instantaneously* to generate a simple idea, by way of dendritic electrical signaling. A dendritic pleat can in fact be compared to a computer's microprocessor (see Ch. 11, p. 212).

Dendritic Pleats and Progression of Memory

Now we will explain the progression of memory – from sensory memory to LTM.

- ***Sensory memory*** initiates the grouping together of the concerned dendrites, because molecular grids that fire together, wire together. This results in the formation of dendritic pleats.
- These dendritic pleats may break down immediately, in which event the person fails to register the perceptions into memory (as happened to our artist with the dinghy in the above example).
- Dendritic pleats may remain together for a short while – in which case they represent memory traces in ***STM*** (like that of the artist who

has taken notice of the mountains, lake, flowers and speeding boat for his painting). See 'Mechanism of durability of memory', below for details of how the dendritic pleats bind together.
- If the dendrites in a dendritic pleat are held together for any extended period of time then ***LTM*** results. See 'Mechanism of durability of memory', below for details.

Memory Recall

How is the stored memory recalled? Three types of memory recollection are possible, as explained below.

Type I

This is a simple type of recall. Here the object or event that is stored in the memory can be recalled when you present the same stimulus to reactivate the memory. In the above example, the impression of the odd chip is stored in memory, and when the same chip is shown it can be recalled from stored memory. Here the memory recall is due to reactivation of the existing dendritic pleats. This excitation is initiated by an external stimulus that matches the stored memory – it gives a predictable and dependable response from the stored memory *instantaneously*.

Type II

Here we recall a memory from the *past* – there is no *present* stimulus. For example, you can recollect the odd chip in your mind long after the odd chip is removed from your sight. There is no external stimulus, but memory is reactivated. The dendritic pleats have ***consolidated*** into a specific pattern that when excited gives you a specific sequence of perceptions for which they are designed (thus forming the idea of an object in question). We do not know exactly what initiates the memory process, whether it is *de novo* or *externally induced* (Ch. 10, p. 200).

Type III

A major chunk of the human memory is of this third type, and this appears somewhat complicated. Here the memory traces may get skewed or swapped. The memory traces of a complex idea are jumbled and rearranged to give rise to new complex ideas. Or sometimes the brain has to 're-synthesize' entirely new ideas by taking other memory traces that were not originally part of the original input. Consider the examples given in Chapter 4 (pp. 89–90), which shows how new ideas may be experienced; for example, we may experience the flavor of a new dish, but this is in fact a mixture of various memory traces from different smell and taste sensations. This sort of memory reconstruction is the most versatile and advanced, as in the case of humans. Consider optical gimmicks, such as the rabbit–duck paradox and the Rorschach ink-blot image in Chapter 9; they are also examples of this type of memory. Human beings are masters in such reconstructions, and sometimes they form new ideas that are so deceptive that they appear as though some totally new ideas are born out of the mind (see our *alien* in Ch. 4). And this is the explanation for weird human perceptions such as *hallucinations, delusions* and *déjà vu* (see below). What directs the mind to construct these complex ideas and what initiates them? These questions are tackled in Chapter 10, as the answers will be vague and uncertain.

Mechanism of Durability of Memory

Now we must answer the following question: how do the dendrites 'stick' together and 'stay' together in the form of dendritic pleats for variable periods?

All animal cells stick together chiefly by means of various transmembrane proteins, and neurons also have these proteins in abundance. In 1970 Gerald Edelman identified certain ***cell adhesion molecules*** (**CAMs**) that were involved in outgrowth and bundling of neurons. The CAMs were found to be responsible for the structural integrity and functioning of the CNS – their deficiency leads to the development of mental retardation and other embryonic defects. Glycoproteins such as ***cadherins*** are also identified, which bind cells more tightly and with a high degree of specificity. Apart from these proteins, the gap junctions (which constitute electrical synapses, see above) may also play a part in holding the dendrites together.

It can be presumed that the CAMs of dendrites interact with each other to bind together. As discussed in Chapter 7, there are several types of molecular interactions that can bind biological molecules – covalent bonds have the strongest interactions, ionic bonds and hydrogen bonds have intermediate forces and the weakest are the van der Waals' forces.

Therefore, it can be stated that CAMs showing the strongest interactions (covalent bonds) cause durable adhesions between the dendrites, which results in LTM, or that CAMs showing the weakest interactions (van der Waals' forces) cause transient adhesions in the dendritic pleats, resulting in sensory memory. It can also be said that CAMs that have intermediate forces result in STM, as shown in Table 8.2.

It can be said that most of the day-to-day human memory is 'engraved' or 'encoded' in the brain in the form of LTM, e.g., the familiar faces we see, the alphabets we have learnt, the oft-used names, or the routine objects we see in and around our houses, etc. All these memories can be recollected *instantaneously* because they are engraved in our memory in the form of *permanent* dendritic pleats.

TABLE 8.2 Chemical Bonds and Durability of Memory

Interactions Between CAMs	Duration of Binding in Dendritic Pleats	Type of Memory
Van der Waals' forces	Transient	Sensory memory
Ionic bonds, hydrogen bonds	Intermediate duration	Short-term memory
Covalent bonds	Durable interactions	Long-term memory

Dendritic Plasticity: Formation of New Memories

Synaptic plasticity (see Ch. 3, p. 67 & Fig. 3.9) is the ability of neurons to connect between themselves and to *change* the connections in response to use or disuse. The name 'synaptic' naturally means that the memory is stored and recollected in the *synapses*; however, *dendrites* are the most versatile structures of the neurons, and in the above sections we can see that dendritic pleats are responsible for memory. Hence, it is more appropriate to use the term ***'dendritic plasticity'***.

We can now say that if we change the order of dendrites in a dendritic pleat it results in a new memory trace – because when the order of generation of primary thoughts changes, it produces a new idea (Ch. 4, p. 82; Fig. 8.4). And it can be guessed that to form a new idea the dendrites have to swap their positions in order to rearrange the sequence of primary thoughts. For this to happen the adhesion molecules which bind dendrites together, such as CAMs, should be able to make or break their bonds readily and rapidly – and this is the reason for the phenomenon of dendritic plasticity.

Consequently, we can presume that CAMs are essential for the development of memory and learning – and this explains why experiments consistently show increased protein synthesis during learning (see below).

Protein Synthesis and Memory

It has long been known that proteins are involved in the mechanism of learning and memory in human beings and higher animals. It is experimentally established that protein synthesis in the brain is accelerated during learning, and it was also proved that if protein synthesis is blocked in the neurons the formation of LTM is abolished. The importance of protein in memory is stressed in Chapter 7. It has been shown in the above sections that proteins are required for the formation of dendritic pleats, and that dendritic plasticity essentially involves protein interactions between the dendrites in the form of CAMs. It is a common observation that the plastic capability of some older people becomes so rigid that it takes a long time for them to form memory traces, and this makes them much more reluctant to learn new things – and this is because of the tardiness of protein synthesis as age advances. This becomes more prominent in people with Alzheimer's disease, where an abnormal protein called *amyloid* is known to accumulate in neurons, which interferes with the formation of normal proteins. It is certain that all these phenomena work by interfering with the formation of normal dendritic pleats.

Evidence for Dendritic Pleats

Researchers at Northwestern University Feinberg School of Medicine (Zelano et al. 2011) reported strong evidence showing that the brain uses predictive coding in the olfactory cortex to generate 'predictive templates' of specific smells. In the moments *before* you stop and smell a rose, it is likely that your brain is already preparing your sensory system for that familiar floral smell. They have suggested that, if somebody asks you to say whether a sample of milk is sour, your brain starts to form predictive templates before sniffing. They have monitored the activity of the brain by fMRI during 'odor search tasks' and confirmed the existence of predictive coding mechanisms in olfaction. Perhaps this is strong evidence for the existence of dendritic pleats coding for memory traces in the human brain.

Experimental Evidence

Memory consolidation takes probably about 5–10 minutes and consolidation is completed after about 1 hour or so – and it has been shown that if protein synthesis is blocked in animals during the acquisition of LTM then the formation of LTM is prevented (Guyton 2008, p. 726). Furthermore, it is documented that if protein synthesis is blocked after about 4 hours, it shows no effect on learning (Barrett et al. 2012, p. 285) – obviously because consolidation is completed and there is no need for new CAMs. Perhaps these experiments stand as good examples for the role of dendritic pleats in LTM.

FORGETFULNESS

What actually is 'forgetting'? Failing to retrieve an event does not mean that this specific event has been forgotten forever. Some of the thoughts once registered as LTM can, theoretically, be retrieved back into memory at any moment in the future. For example, a man forgets his old schoolmate and is unable to recollect his face or his name after a certain period – the memory of this schoolmate *appears* to be 'totally washed out' of his memory. In this type of memory it cannot be said that it is erased *completely* – the memory may reappear at any future moment if the correct stimulus is applied, such as a specific incident (see Ch. 10, p. 200). Perhaps this means that, once formed, memory remains stored in the brain forever as memory traces.

Paradoxically, the molecular grid model offers an opposing view of forgetfulness. It can be argued that memory can be erased *completely*. As we have seen, the key to memory is the formation of dendritic pleats, and these pleats were bound down by molecular interactions, as shown above. This shows that if the dendritic pleats are somehow 'disbanded' or broken down, then those memory traces can be *totally forgotten*. Thus, it can be said that a memory can be *totally* and *permanently erased*. The disruption of dendritic pleats can occur by any of the following means:

- By natural means (e.g., physiological remodeling of childhood memories, aging, natural changes, etc.)
- By physical means such as brain injuries (e.g., post-traumatic amnesia – here the dendritic pleats are physically broken)
- By chemical means (e.g., chronic use of drugs like alcohol, sedatives, narcotics)
- By some psychological means (e.g., hypnotherapy)
- By certain neurodegenerative disorders because of accumulation of 'bad' protein (e.g., Alzheimer's disease)
- By some other forces, possibly electromagnetic (e.g., the theory that exposure to microwaves in cell phone users lead to cognitive problems).

INTELLIGENCE

Human intelligence is an intricate affair, for which the most fundamental requirement is memory. Overall performance of the brain depends on the quality of perception – for example, the higher the quality of the things we see, hear or feel, the greater the knowledge we acquire from them. The quality of vision, sound, touch, etc., improves with the ***number*** and ***quality*** of primary thoughts in that perception; for example, the image resolution of a 14-megapixel camera is far superior to a 3.2-megapixel camera, because the greater number of pixels results in a better-quality image. In the same way, if a perception comprises more primary thoughts, then that perception is sharper. The quality of primary thoughts also dictates the quality of perception – for example, a person with blue-yellow color blindness is deficient in perceiving the color blue and perceives blue as green, thus his appreciation of a blue-colored object is compromised to a great extent. Hence it can be said that the greater the number of primary thoughts with high quality improves the overall performance of a person.

The molecular grid model offers two interdependent factors that can determine human intelligence: (1) proteins in the molecular grids; (2) lipids in the molecular grids.

Proteins and Intelligence

This factor has already been adequately stressed. As the quality of primary thoughts dictates the quality of perception, and consequently directs the level of intelligence, it is a simple matter to say that the quality of proteins decides the level of intelligence. And, it can be inferred that poor nutrition during the formative years of childhood interferes with protein synthesis, and this may compromise the intelligence of the child.

Lipids and Intelligence

Now we will come to another important but inevitable conclusion. Chapter 7 shows that proteins in a microdomain are recruited by the lipids; this shows that the nature of lipids in the neuronal membrane directs the type of proteins in a molecular grid. Thus, if the dendritic membrane is deficient in certain essential fats the protein recruitment may also become faulty – consequently this may result in a defective memory, thus leading to less intelligence. As a case in point we take DHA and see how this affects the performance of an individual.

DHA is an *omega-3 fatty acid* found abundantly in the brain and other tissues. Some estimates show that DHA accounts for almost 50% of neuronal membrane's weight (Singh 2005). Numerous scientific studies show a definite connection between brain development in children and DHA intake. A study from the University of Alberta shows that 'the memory cells in the hippocampus could communicate better with each other and relay messages better when the DHA levels in that region of brain were higher'. It is an anecdotal supposition that a diet rich in fish (a good source of DHA!) improves mental ability and the intelligence of an individual. The molecular grid model offers an explanation for this phenomenon.

It is possible that DHA is concentrated in the lipid rafts of the neurons and helps in maintaining the integrity of molecular grids perhaps by reducing the *hydrophobic interactions* (Ch. 7, p. 134) between lipids and proteins. This helps in recruiting more proteins into a molecular grid. Molecular grids may fall apart when DHA is deficient in them, thus interfering with their capacity to generate primary thoughts, and obviously this interferes with the ability of perception itself. Thus it can be said that a brain deficient in DHA (or some other types of essential fats) can form fewer memory traces, and consequently this reduces the ability of the cortex to hold memory and thus compromises the level of intelligence.

SOME WEIRD BUT EXPLAINABLE PHENOMENA

Below are a few phenomena that have defied proper explanation over the ages. The molecular grid model tries to explain them, at least in part.

Are There Memory Traces at Birth? (Memory Traces sans Memories)

As explained in the above sections, the dendritic pleats are formed when an incoming sensory stimulus reaches the dendrites and these dendritic pleats give rise to memory traces. This means that when there are no sensory stimuli there will be no memory traces. We can naturally presume that there are no memory traces (or dendritic pleats) at birth, because there are no sensory inputs yet, at least not many. Or in other words, the mind is a *tabula rasa* – a blank sheet – at birth. But this presumption can be contested in the following discussion.

It appears that some basic memory traces are formed even at birth, even without any sensory experience. This can be better understood if we take animals as an example. There are many animal species that give birth to their offspring in a fully developed state. Many reptiles and a few birds hatch their eggs and as soon as the hatchlings come out they lead independent lives without being dependent on their parents for food, movement or predation. These are called *precocial* species (in contrast to *altricial* species, which are born 'helpless' and depend on their parents for variable periods). A common example of precocial species is the domestic duck – as soon as they are born they open their eyes and run in search of food, they can swim and they can identify their food by smell, identify their enemies, etc. These are all survival instincts that are formed prior to sensory experience. A human child is essentially altricial, and thus has limited ability to help itself soon after birth. However, even human babies may have a store of memory traces at birth, as evidenced by some strange scientific reports that babies in the womb have REM sleep and dreams even before birth, and that intrauterine babies can enjoy music in the womb, etc!

How are these memory traces possible without sensory stimuli? These phenomena may be explained by considering that dendritic pleats are formed before birth with the help of genes. It has been shown that protein synthesis is essential for the formation of dendritic pleats (see above), thus it can be presumed that the genetic mechanism has already directed certain proteins to create these dendritic pleats, but without the necessity for prior sensory experience. This means that our brains were already preloaded with memory traces – the mind was *never* a *tabula rasa* but is already 'smeared' by basic memory traces even before we are born. Is REM sleep in the womb evidence of the formation of such new dendritic pleats and the *consolidation* of memory taking place?

Child Prodigy

We are familiar with child prodigies performing musical feats (e.g. Mozart) or math geniuses, such as Srinivasa Ramanujan, or born artists (e.g. Pablo Picasso) – their brains appear to be somehow 'channelized' to perform these complicated actions, even at an early age. There are examples of some children excelling in other forms of art, and it is astonishing that they seem to have been born with these skills. In some instances, we can also observe that a musician's child excels in music, a linguist's child excels in languages, etc. All these extraordinary abilities are possible because the dendritic pleats have already formed owing to the existing special proteins synthesized by genes inherited from the parents – thus the primary thoughts can readily assemble into ideas without much difficulty, and this facilitates memory formation with relative ease. This must be the reason for their precocious proficiency in their respective arts. Their skills are 'half-finished' before they are born, and it takes only a minimum stimulus for them to surface. No wonder, we can see now that a musician's child can become proficient in music without much training, and an artist's child can be a born artist.

Hallucinations and *Déjà vu*

The bizarre experiences of hallucinations, delusions and *déjà vu* (Ch. 3, p. 74) may also be explained by considering that dendritic pleats and memory traces have formed prior to experience in some assorted fashion. It is possible that these dendritic pleats are somehow swapped or skewed, leading to the production of distorted memory traces (see above) – hence their abnormal perception.

Now, we will go to Part III of the book. Hitherto, in Chapters 4 to 8, we have discussed ***thought as an electrochemical process***, arising out of a stimulus traveling to the neurons and exciting them electrochemically and generating primary thoughts, but nowhere in these chapters have we dealt with the ***exact*** nature of thought itself. It has been pointed out in Chapter 7 that the exact nature of thought remains elusive.

The next two chapters will deal with the lesser known aspects of thought. These issues are very complex and fall more into the domain of philosophy than science; however, we will deal with them scientifically to unravel the mysterious nature of thought. We will first see what constitutes ***mind*** and ***consciousness*** in Chapter 9 and tackle the problem of the central executive. In Chapter 10, we will treat thought as a ***metaphysical entity*** and examine the relationship of mind to matter, which is an age-old perplexity.

PART III

THE EVOLUTION OF HUMAN MIND

CHAPTER

9

Mind and Consciousness

OVERVIEW

Chapters 4–8 examined ***thought as an electrochemical phenomenon,*** generated by the neurons at cellular level. Throughout these chapters we have seen how a stimulus travels up to the neurons and excites them electrochemically, generating primary thoughts and ideas. Chapters 9 and 10 will touch upon some of the vague aspects of human thought that have defied strict scientific examination from time immemorial, and hence have fallen into the gray areas of science.

The fundamental understanding of ***mind*** and ***consciousness*** is one such aspect. Here in Chapter 9 we undertake a scientific study of these concepts, and the reader can see that the following discussion is entirely bereft of vagueness and obfuscation, which is so common when the abstract topic of mind and consciousness is discussed. In this chapter we will also discuss how the human mind has progressed to achieve intelligence by examining the versatile ***central executive*** (CE) that is so characteristic of human beings.

The other aspect of human thought that has eluded proper scientific examination is the exact nature of thought itself. In Chapter 10 we will examine ***thought as a metaphysical phenomenon*** and we will see how the mind and matter problem can be studied by taking the molecular-grid model into consideration.

EVOLUTION OF THE HUMAN MIND

The human mind can be considered as a complex set of cognitive abilities, such as perception, thinking, consciousness and reasoning, and most of these are dependent on higher intellectual functions of the brain. However, there is a great deal of vagueness in defining the human mind,

The Biology of Thought.
DOI: http://dx.doi.org/10.1016/B978-0-12-800900-0.00009-9

and this is understandably because of its all-pervasive nature. This chapter examines the concept of mind scientifically and arrives at important conclusions.

At the outset it must be understood that mind is not a phenomenon peculiar to human beings – rather, the human mind has gradually evolved across generations from the lower organisms. A detailed scientific study is undertaken below to show this gradual evolution of mind from animals to man.

Perception as Awareness

All organisms have a basic *desire* to live and, in order to cope with their environment, all life forms on earth have to perform the following ***four basic survival functions***:

1. They have to procure food
2. They have to protect themselves from predators
3. They have to grow in size
4. They have to procreate.

To perform each of the above functions, all life forms have to *sense* the external world – or, in other words, they must receive stimuli from the external world. For example, an organism has to sense the presence of food in its surroundings in order to reach for it. Likewise, in order to protect themselves, organisms need to sense their predators; and in order to grow and procreate they must sense favorable conditions in their vicinity. What are the mechanisms that various organisms employ to perform these acts?

In the case of *microorganisms*, they employ various techniques such as chemotaxis, mechanotaxis, etc., to *'sense'* their external environment and carry out their survival functions. For example, amoebae have to sense the presence of food in order to move their pseudopodia towards it, and they have to sense the presence of a favorable environment before they divide – and if they sense unfavorable conditions they tend to form spores. *Plants* employ phototaxis, geotaxis and other such mechanisms for their survival and propagation; for example, a creeper in the garden has to sense its surroundings in order to creep towards the source of sunlight. Not only for food, but all other survival acts have a sense of the environment as a primary requirement – a vine tendril senses the presence of a beam as a support, reaches out for it and wraps around it (*thigmotropism*).

In other words, all organisms must be ***aware*** of their surroundings in order to execute their survival functions (Fig. 9.1). This means that ***awareness*** is the characteristic feature of all life forms. Thus, we can make the following statement:

Awareness is sensing the environment, and all life forms possess awareness.

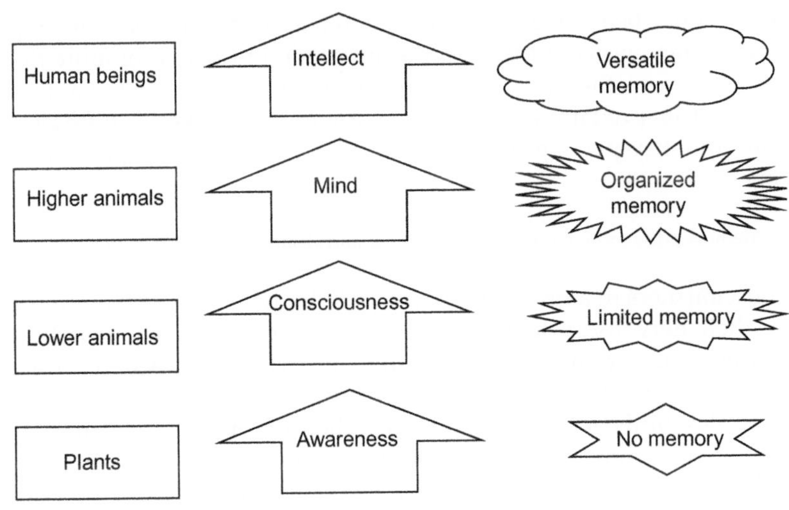

FIGURE 9.1 Awareness–consciousness–mind–intellect.

Origin of Consciousness and Mind

Now we will see how this awareness in the case of animals advances into ***consciousness***. All animals also need to demonstrate the above survival functions, and thus all animals also must be aware of their surroundings in order to survive. However, all animals (from invertebrates such as worms, to insects, reptiles, birds and mammals) become aware of their external environment by way of senses such as smell, vision, sound, touch, etc. This advanced form of awareness is possible only when they have a ***nervous system*** to assist them, and thus all animals possess some sort of nervous system. By using neural mechanisms, animals can ***perceive*** these sensations and thus have become capable of moving around in space more effectively and performing all four basic functions. Thus, we can say that:

> Perception is awareness.

The characteristic feature of a nervous system is not only to perceive the senses but also to hold these perceptions for at least a brief period, i.e., to form ***memory*** (Fig. 9.1). Thus, all these animals are not only aware of their surroundings, but are capable of holding this awareness in memory for variable periods, and this helps them to navigate in their surroundings efficiently – even a snail needs some slight memory to move around in space. This sort of awareness stored in the nervous system as memory is called ***consciousness*** (Fig. 9.1). The animals now have become ***conscious*** of their surroundings (*not* just being *aware* of their surroundings, like plants). A busy bee hovers over flowers in a garden until they are empty of their nectar, and could fly to a distant garden in search of more flowers, which means that it is *conscious* of the next possible source; a deer is *conscious* of its predators in the jungle; a mother monkey is *conscious* of its baby monkey playing nearby, and so on. Hence we can pass the following statement:

Memory is the cornerstone of consciousness, and all animals possess consciousness.

Consciousness in animals progresses into higher levels as the complexity of their nervous system increases. The lower animals (invertebrates and insects) have only a very short memory which works only to serve their immediate survival instincts. As animals become more advanced (e.g. reptiles, birds and lower mammals) their brain's neural connections become increasingly complex and the consciousness progresses to provide them with a longer memory and, more importantly, in a better organized fashion. Thus, as animals ascend in their phylogeny they become more capable of organizing their memory to perform the four survival functions – an 'eyeless' earthworm can only move around in space and cannot have any further memory; an insect such as a mosquito has a little more memory to repeat its movements over and over; birds such as parrots can do far better in memorizing external events; a mouse in a maze or a monkey in a predicament can store more details in their memory and overcome all hurdles with their skills. Thus, all these animals possess consciousness in increasing degrees as they become more advanced.

Origin of Mind

The nervous system in the more advanced mammals, such as elephants, dogs, tigers, etc. (Table 9.1), is much more complex, and their consciousness is much more advanced and can store much more memory. In fact, the memory in these advanced animals takes the form of a solid long-term memory (LTM), which can be used to perform certain intelligent tasks. This advanced form of consciousness is the ***mind*** (Fig. 9.1). Thus, elephants are not only conscious but they have minds of their own, as is the case with your pet dog. As we can see, the simple survival function of awareness in lower life forms has transformed into a complex mind in higher animals, and reached its zenith in human beings. Now we will consider the human mind itself.

The Human Mind

In human beings consciousness has fully developed into *an intelligent and reasoning mind* (Fig. 9.1), with a good chunk of it working in a very organized manner to form working memory (Ch. 3, p. 58 & Ch. 8, p. 147). We will dig a little deeper to see what exactly the mind is composed of.

TABLE 9.1 Comparative Analysis of Development in Animals and Man

	Worms and Insects	**Fish, Reptiles and Birds**	**Advanced Mammals (elephants, apes)**	**Humans**
Will to survive	Only will to survive	Will to survive present	Will to survive present	Will to survive present
Volition	No volition	Limited	Greater degree	Fully developed
Memory	Only sensory memory	STM and some LTM	STM and limited LTM	Full LTM
Awareness	Only aware of surroundings	Awareness present	Awareness present	Awareness present
Consciousness	Primitive	Limited	Greater degree	Advanced
Development of CNS	Primitive no cortex	Limited cortex	Cortex of variable size	Expansive cortex
Free will	No	No	? Doubtful	? Advanced
Central executive	Least developed	Developed to some extent	Very well developed	Advanced

We have shown that awareness in the lower organisms transforms itself into an advanced form called consciousness in animals, which then progresses into a fully developed consciousness, called mind, in humans. This means that at the roots of mind there is awareness, i.e., mind takes its origin from awareness. And because awareness is perception (see above), we can reasonably say that the fundamental constituent of mind is perception. We saw in Chapter 4 that perception manifests at the level of neurons in the brain as primary thoughts. All this means is that primary thoughts are the essential constituents of mind. Considering the above facts, it can be extrapolated that for the mind to exist it must constantly perceive signals from the surrounding environment. Thus, it can be stated as follows:

Mind is the product of continuous streaming of primary thoughts.

We can appreciate this phenomenon better if we follow some examples given below.

1. **The anechoic chamber**. *Orfield Laboratories* in South Minneapolis, USA, have constructed a soundproof chamber to test the sound qualities of various products (such as cell phones, hearing aids, bikes etc.), and this *anechoic chamber* is proclaimed by the *Guinness Book of World Records* to be the *'quietest place on the earth'*. Human volunteers

have tried to experience what it would be like in such a chamber alone for some time. As an experiment, volunteers sat in stagnant air, had to face a corner with no light, no odor and, of course, no sound at all. It was observed that after spending about half an hour in such a bleak and dark cubicle, the volunteers became disoriented and eventually they experienced visual and auditory hallucinations. Finally, it was proved to be an utterly intolerable situation. This shows that for the mind to function properly, a minimum input of sensations from the external environment is required.

2. **Perceptual isolation.** It is a lesser known fact that severe deprivation of sensory stimuli (*'perceptual isolation'*) has been used as a torture technique by the armed forces, subjecting victims to a near total deprivation of external stimuli, and this leads to extreme anxiety, hallucinations and irrational thoughts in the victims (the so-called *'ganzfeld effect'*); however, the use of such inhuman techniques is no longer accepted by NATO. It is also a common observation that total silence sometimes leads to utter boredom (*ennui*). This feature also demonstrates that the human mind goes awry if it is deprived of sensory input for long periods.
3. **'REST'.** However, such sensory deprivation is used in a controlled manner by some psychologists to treat various addictions such as smoking, alcoholism, drugs, etc. One such technique is REST (restricted environmental stimulation therapy), which has had some promising results (Borrie, 2002) in treating addictions. The use of biofeedback techniques also offers some benefit in treating conditions like migraine. Perhaps the principle behind all these techniques lies in the fact that controlled deprivation of senses subtly alters the thinking mind and consequently modulates the conscious appreciation of 'self'.
4. Current research shows that there are constant ***background gamma waves*** (60–140 Hz) generated by ill-defined centers in the brain called ***default-mode network***, which keep on working even when we are totally idle, as the *'the brain hates emptiness and never stays idle even for a tenth of a second'* (Ossandón et al., 2011). This background awareness is probably the result of a continuous streaming of primary thoughts and ideas across the mind.
5. **Meditation.** Meditation, as practiced in certain types of yoga, is chiefly based on the restriction of sensory inputs into the brain, which calms down the mind and cures certain illnesses.

This shows that the human mind needs a minimum (but constant) input of stimuli from the external world – this is because mind takes deep roots in awareness, and awareness is nothing but perception. This again means that the primary thoughts and ideas we make from the external world become the general background of our conscious mind, without which the mind goes bankrupt.

EVOLUTION OF THE CENTRAL EXECUTIVE

In this section we will see how the human mind has progressed to develop intelligence and intellect that no other animal is endowed with. The most outstanding ability of the human mind is to put its LTM into tremendous order and perfect categorization, which helps man not only to remember the past accurately but to extrapolate memory into future predictions (see *semantic memory* in Ch. 3). Obviously the mind needs a versatile executive system to categorize events and to memorize them in a coherent and rational order. This agent, or this executive system, is called the ***central executive***, which works constantly in the background of the human mind. We have seen something of this CE in Chapter 3 in relation to working memory, and now we will see what exactly is this CE and how it has evolved in human beings.

We can define the ***CE*** as an agent which has the ability to put primary thoughts and ideas in order, so as to produce a meaningful image of the external world and store it in memory for an extended period of time so as to allow its retrieval in future in a rational sequence. This rational memory is the cornerstone of human intelligence.

As an example, in order to say that you are looking at a real rose flower (not a paper rose) you have to not only *look* at the rose, but you have to *smell* for its fragrance and *touch* the flower for its characteristic feel. Only if all these primary thoughts received from a rose, regarding its color, shape, smell, touch, etc., are arranged in an order and assembled into a complex idea, can we appreciate it as a rose (Fig. 9.2) – if the arrangement is jumbled, perhaps as in the case of an insane person, it is no longer perceived as a rose. Now consider this – after examining the rose by its look, smell and touch you have ascertained that what you have there is a *paper rose*, not a real one, then you may proceed to entertain other rational thoughts about it – for instance, you may now say that an origami artist has created this paper rose and that the original art of origami has its roots in Japan. All these rational thoughts are orchestrated by the CE, which arranges these complex ideas in an organized manner to make perfect sense of the things we observe out in the external world and to ascertain their implications. Once these complex ideas are formed they are stored in LTM, which can then be recalled at appropriate moments; and over a period of time many such coherent ideas build up human intelligence.

The characteristics of this 'internal manager' are not yet known, and its existence itself is barely tangible. But it must be realized that this CE sits at the top and bosses around all of our thoughts and consequently our

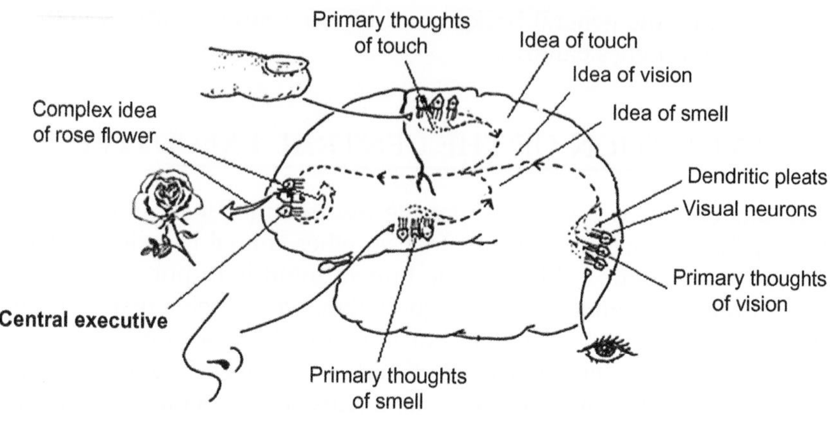

FIGURE 9.2 The central executive.

activities. However, it can be said that the CE differs from individual to individual, and each one of us has a characteristic CE within us.

The Unique Central Executive!

The human mind receives sensations from the external world and these are interpreted as perceptions. The received sensations may be similar, but the perceptions in the mind vary in different individuals (Ch. 4, p. 82 & Ch. 8, p. 158). Thus, each one of us may interpret a similar set of sensations in a different manner, because we tend to analyze the received data according to our own preferences, inclinations and experiences – as directed by our own CE. The following examples will better clarify this point:

Example 1. See the ***optical illusion*** in Figure 9.3: you may see the impression of a duck, and another person standing by your side may perceive the image of a rabbit. Or even you yourself may see a rabbit sometimes and a duck at some other time. What is it that directs our thoughts? What weaves our primary thoughts together to give different complex ideas? It can be said that the CE, having dominance over the mind, is managing the assembly of primary thoughts into a characteristic pattern of ideas for us. And this CE goes by so many names – instinct, intuition, preference, interest etc.

What you have actually received is numerous dots of black ink, but they have assembled into different sequences according to our preferences or inclinations. We can also play around with our CE by watching wispy clouds in the afternoon sky and finding their resemblance to common shapes (***cloud graffiti***) – a cloud may look like a deadly lion to you, but your friend may perceive it as merely cotton candy.

FIGURE 9.3 Rabbit–duck paradox.

FIGURE 9.4 Rorschach ink-blot image.

Example 2. In the same way, how do you interpret the ***mirror-image design*** shown in Figure 9.4? This sort of random image can be produced by simply spilling some ink over a white sheet, folding it in the center and pressing it gently. Can we say that this is an image of two witches performing some ominous sorcery? Or should we say that this is a picture of a mad wolf with a pair of angry eyes, two upright ears and a big snout with dribbling saliva? Or we may see dilapidated huts, upturned horses, bears or butterflies in the image. Who manages this assembly of thoughts? An internal manager, surely!

Example 3. Now consider word jumbles. It can be demonstrated that we can automatically put a jumbled-lettered word into a readable sequence, so that it can be read normally. For example, if you casually read the sentence in Figure 9.5, you will find it a very ordinary sentence. But only if you carefully look at each individual

THE PAFICIC OCAEN IS THE LARGERST OCAEN

WHLIE THE ACRTIC OCAEN IS THE SMALLAST

AISA IS THE LARGERST CONITENENT WHLIE

AUTRSALIA IS THE SMALLAST

FIGURE 9.5 Jumbled words.

word, you will notice the jumbling (whereas a diligent proofreader may catch the mistakes right away). How did we perceive the correct words effortlessly in place of the jumbled words in our mental eyes? Who managed this?

All these examples show that the CE is unique to every individual, and operates in the background of his/her mind, and moreover, we can see that the CE can be trained to perform the tasks an individual desires to undertake. Thus, the overall intellectual performance of an individual depends on the make-up of his/her CE. We can now pass on the following statement:

> The central executive is the intellect behind the human mind.

Where did this CE originate? To understand the evolution of the CE in human beings, we have to first examine the phenomena of ***will to survive*** and ***volition*** in lower animals. A discussion of will and volition will also pay unforeseen dividends by providing answers to seemingly unrelated questions, such as what separates man from beast, why we behave as we do, and why are we so emotional, as discussed below. First we will see what is will, and what is volition.

Will and Volition

Will to survive (or simply '***will***') is the basic urge of all organisms to live. It is the fundamental property of life, which is manifested perpetually in every organism. We will now show that the will to survive can be distinguished from another phenomenon called volition. ***Volition*** is the capability of an organism to perform a voluntary action, and it is the organism's advanced property. The distinction between will and volition becomes obvious with some explanation.

All organisms have a desire to live, and consequently, all life forms have to perform the four basic functions (described at the beginning of this chapter) for them to cope with their surroundings. These are the survival instincts. This is the *will to survive* and it is the essence of life. Most of the actions of the lower animals can be related to the will to survive (see the example below). But then, what is *volition*? We can say an organism possesses volition if it is able to demonstrate any action that is *not directly* related to any of these four survival mechanisms. The following discussion makes the point clear.

A housefly constantly surveys its surroundings for food or its predators or its mate – it demonstrates all its will to survive to the fullest possible extent – and it can be shown that the housefly *never* diverges from these survival acts. For example, as soon as you leave a piece of cake out in the open a housefly may hover over it and sit on it to savor the cake's sugar – this is the will to procure food. The housefly sucks in the sugar, not only to obtain energy but also to grow in size – this is the will to grow. The housefly also seeks a mate – this is the will to procreate. It surely has a few spontaneous or deliberate actions – such as when it tries to quickly fly away and escape a hit from your flyswatter – which is a survival instinct to avoid danger to life. In other words, it is almost incapable of an action that is not related to one of the four survival mechanisms mentioned above. You cannot expect a housefly just to step out and go for a picnic, or to dance around a garden graciously but *purposelessly*.

In the same way, a fish in a pond always goes in search of its food, defends against its predators and enemies, and seeks copulation to maintain its progeny. All amphibians (e.g., toads) and reptiles (e.g., snakes) also perform acts that are strictly directed towards the maintenance of these basic survival instincts. Their actions are always related to one of the survival mechanisms mentioned above. In other words, all these actions further promote the purpose of the will to survive.

Many birds (such as sparrows) also demonstrate only these survival instincts, but some advanced birds, such as parrots, can perform acts like mimicry and demonstrate other learning abilities that are not particularly essential for their survival. This means that, although their will to survive is well preserved, they can also perform certain activities that are less connected with these survival instincts. Mammals are phylogenetically advanced animals; many of them, such as tigers, bears, elephants and dogs, can perform acts far removed from the survival instincts. They can be seen to play for 'the sake of fun' with no essential purpose useful for survival. Basic instincts such as remembering their food and water source, engaging with their mates and keeping an eye on their enemies, are done as efficiently as in the lower animals; but they also do certain things that are apparently not directly related to survival – they can demonstrate 'play' behavior, express certain emotions like pleasure, and can even become self-aware to some extent. You can play with your pet dog and see how its volition works. This shows that these advanced animals can perform *voluntary actions* better than primitive animals. They can be said to have a higher degrees of ***volition***. This means that they can *choose* to act or not to act *voluntarily* (volition = voluntary action).

As the phylogeny advances, the behavior of animals becomes more complicated and far more detached from survival instincts. When we take the example of primates, not only do they perform all the acts of survival (the four features mentioned above), but in addition they do much more – they have the ability to play more freely at will, can display emotions better by using facial expressions and gesticulations, and can communicate more clearly with humans – in short, their volition is far more advanced. When you look at a chimpanzee (chimps are the closest relatives of humans!) you notice that they can play, laugh, make some facial expressions to portray emotions better than other apes, they can even make and use tools, etc. All these actions are highly voluntary and they can choose either to perform or not to perform them.

And lastly, there is nothing to write about humans – they display behavior so far removed from the survival instinct. They can smile, grin, laugh, grieve, play pranks on others, sulk, cheat, love and hate, display morality, think of good and bad – and what not! All these actions are not directly connected with survival mechanisms. This is volition expressed to its highest degree, i.e. the capability for voluntary action has reached its zenith in the case of human beings (see 'Volition, memory and central executive', below).

Characteristics of Will

It must be stressed that the will to survive remains the same in all organisms; it is expressed as vividly in the case of a housefly or a fish as in the case of an elephant or a man. After all, survival is survival for any organism (compare with 'volition', below). The will does not need any nervous system to express itself – it exists in plants and microorganisms and in animals as well, otherwise how would they cope with their surroundings (see below)? As long as life exists in an organism there is a will to survive, which fights constantly for life – the will to survive would never give up (see 'Can the will be conquered?', below). Will does not need rest or sleep – it does not get tired – it works in the background at all times. There is no boredom for the will (compare volition, below).

Characteristics of Volition

Volition, however, differs vastly in its expression from one organism to the next. The phenomenon of volition depends on the development of the nervous system – and it can be observed that as the complexity of the nervous system increases, the exhibition of voluntary actions also increases ('Volition, memory and central executive, below). Volition needs rest and sleep, and it may get tired or give up the struggle at some stage. Boredom (or *ennui*) is the characteristic of voluntary acts – as the nervous system attains supreme volition (as in humans) boredom also becomes prominent (Durant, 2006, p. 422). This is perhaps the cause of the mental exhaustion we feel at the end of strenuous mental work – during this period we would have used volition to its fullest extent, and soon we would feel 'burnt-out'!

As the complexity of CNS increases, the exhibition of volition improves.

Volition is an *epiphenomenon* – which means that it is added to the already existing will – and they stay together at all times. However, one of them (either will or volition) becomes more prominent than the other at a given time. As an example, a group of tourists exhibit their volition to their full extent when they are having fun and frolics at a beach resort – they surf in the water, chase each other, play mischief and what not, all with their exhilarating powers of volition. A sudden prank call of a shark in the ocean sends them all to exhibit their will to survive more than volition – and *involuntarily* all of them try to come ashore, screaming for help, running for their lives and causing all sorts of pandemonium. This sort of mayhem caused by people is prominently seen during emergencies such as a house fire or floods and hurricanes. In all these situations the basic animal instinct (= will to survive) surfaces to the fore and results in these involuntary, sometimes irrational, steps to look after the safety of the self.

The Will in Plants and Microorganisms

After the above discussion, we may have legitimate doubts – what about other living beings, such as plants? How can we say that they can perform the acts of survival – we have seen animals demonstrating only the will to survive? Plants are living organisms that are generally immobile and do not exhibit visible movements on such a time-scale as that of most animals. They procure food from sunlight, grow, protect themselves, and have the ability to procreate. We need to investigate now whether they also possess some *urge* to perform the four survival instincts described at the beginning of this chapter – if they have an urge to sustain themselves in their surroundings, then we can say that they too have a will to survive.

In a childhood experiment we would have seen that when we isolate a plant in a dark box with a hole in it, the plant exhibits phototaxis by bending towards the light source from the hole. This means that plants have a 'desire' to procure food and live. In the same way they exhibit various movements (though slowly), such as *hydrotropism* (growth of roots towards water), *geotropism* (growth of roots downwards) etc. – all of these can be related to the will to survive. Perhaps the best example of their rapid, animal-like action is the movements in insectivorous plants, such as the Venus flytrap, or sensitive plants such as *Mimosa pudica* ('touch-me-not').

It can be said that plants also exhibit an urge to grow (when given proper conditions), spread across land and procreate. Most plants exhibit various mechanisms by which they become dormant for variable periods if their living conditions turn unfavorable. Plants have devised various ingenious contrivances for cross-pollination and seed dispersal to distant areas – this is an urge to increase the vigor of the species and to spread, all of which are also related to the will to survive. We will now see how they try to protect themselves from animals and pests: some develop sharp thorns, some emanate a strong unpleasant odor to ward off animals, some possess poisons that are toxic to animals, etc.

Let us now briefly touch upon the lives of bacteria, protista (amoebae, algae) and fungi. These microorganisms demonstrate various movements in response to external stimuli, such as *chemotaxis* (e.g. to sense food in their surroundings), *mechanotaxis* (e.g. movement away from shear forces), etc. Most of these microorganisms are capable of living in a harsh environment because they are more able to adapt. Adaptation is a survival technique. All these organisms move around to find food; fight to protect themselves from predators in their own microscopic way; tend to grow and to procreate against many of nature's odds. When the conditions are conducive these organisms procreate; when conditions are less favorable they sporulate. All these mechanisms are expressions of the urge to live, and they can therefore be said to have a will to survive.

To conclude, we may say that plants and lower organisms also exhibit a will to survive as vehemently as animals do, but their actions are not immediately visible because of their slow response to stimuli. However, obviously, they have no volition, as they can only act and react to fulfill their basic survival needs. Thus, the phenomenon of volition is completely absent in plants and microorganisms owing to the total lack of a nervous system (see 'Origin of consciousness and mind', above, & Fig. 9.1). Hence, we can say this:

> Plants and microbes can only demonstrate will to survive.
> They have no consciousness.

Can the Will Be Conquered?

We have learnt from the above discussion that the will to survive is the fundamental attribute of all organisms, and this survival instinct cannot be dominated by anything – it may lie dormant, but is present all the time and surfaces in the event of any danger (see 'Characteristics of volition', above). It tries to override all other *ad hoc* actions taken by the organism, and finally tries to keep the organism alive (see 'Suicide', below). A lower organism displays the will to survive with the same vigor as a higher animal does – all organisms fight for their lives, but their fighting modes are different. Each individual organism strives to keep its 'self' alive at all costs. Finally, it can be said that *survival of self* is of paramount importance to all organisms: the will to survive always wins over volition and comes to the fore to make the organism survive. However, we will study the human mind later in this chapter and see how volition in humans may sometimes conquer the will to survive.

The Phenomenon of Reproduction

The presence of will to survive became apparent from the above discussion, but there is a little mischief attached to it – a *will to live* is ***always*** accompanied by a will to live ***forever***, a will to survive for *eternity*. Each individual of any species strives to conquer death, by *any* means, and if left to their will they continue to live forever. But Mother Nature understands the inevitability of death. Thus, the will takes a surreptitious course to win over the phenomenon of death by devising another phenomenon called reproduction (Durant 2006, pp. 412–419). By the process of reproduction the parents identify themselves in their offspring. This is the reason why every animal demonstrates sexual vigor so prominently – they can see themselves in their offspring, and by this they can sustain their race forever and ever. Nature also supports reproduction in a species more readily than an individual organism's survival, because in the long run the former works out to be more beneficial than the latter.

VOLITION, MEMORY AND CENTRAL EXECUTIVE

Now we will examine volition in some detail and see how volition is related to the memory and how it progresses to form a fully fledged central executive in humans. Let us recapitulate the progression of memory from Chapter 8 and briefly see how sensory memory is processed into short-term memory (STM) and later into LTM. Our artist in that example was looking at the scenery and had taken notice of the mountains, lake, flowery bushes, speeding boat, etc., but had failed to notice a dinghy, because it was lying in one nondescript corner in the lake. Sensory inputs from *all* the details of the scene were carried to the brain, irrespective of their being registered in the conscious mind or not – this is ***sensory memory***. Those details that have enjoyed the artist's conscious appreciation (the mountains, lake, flowers etc.) are said to be incorporated into ***STM***. The artist may remember the scenery and his painting for a variable period of time, and this is ***LTM*** (see Fig. 3.4). We will analyze this in a different light now.

Memory and Central Executive

First of all, how do we bring only some sensory inputs (such as mountains, lake, flowers, etc.) into our conscious appreciation (STM), and push other sensory inputs (like the dinghy) out of memory? What induces us to remember certain things for a long time, and to forget certain others sooner? It can be readily seen that memory is the product of our interests, preferences, predilections and inclinations. We have seen that the CE sits at the top and manages the primary thoughts and ideas. Now we can say that the CE also manages these interests and inclinations – if the CE directs the brain to ignore the memory it is ignored and the memory vanishes 'into thin air'; if the CE directs the brain to register an external event in memory transiently it becomes STM; and if the CE instructs the mind to remember the thing or event it can be stored for days, weeks, months or for life, and can perhaps be promptly recalled on demand (Fig. 3.4).

Now we will see how memory is related to volition. To perform actions related to will to survive, no memory is required; hence, plants and other lower organisms can exhibit the will to survive even when they have no memory at all (as discussed above) – phototaxis or geotaxis need no CE – they are performed instinctively, without using the discretionary powers of volition. On the other hand, to perform an act of volition some sort of memory is necessary – the better the memory, the better the performance of voluntary acts. A parrot needs some memory to recite what it has been taught; a mouse needs more memory in order to repeatedly negotiate a hurdle; for a dog to follow its master, it needs a much more complex memory to remember the rewards it gets by this act; a monkey needs some organized memory to play a trick; and a student needs a superb memory to copy notes from the blackboard. Einstein had to have an extraordinary memory to succeed in his hypothetical equations, to enable him to finally see the relationship between all the mathematical equations, which none in the world had seen before (and he had to actually *forget* the trivial things in life to accommodate his lofty ideas). Thus, it becomes obvious that memory is a cornerstone in the development of human intelligence. Hence the following statements can be made:

Central executive is the cornerstone in the formation of memory.
And, the secret to enhancing intelligence is to enhance memory.

Evolution of Central Executive

Now we will analyze memory in the animal kingdom – starting with invertebrates and insects to birds, mammals and men – and this analysis will help in understanding the evolution of the CE in human beings (see Table 9.1, which helps in the analysis). We have seen that perception manifesting as awareness is a fundamental phenomenon of all life forms, and it has progressed into consciousness in animals (see above). An 'eye-less' earthworm has to perceive objects in order to navigate in its surroundings, which means that it can generate primary thoughts from the external world so that it can form an 'image' of the surrounding objects and store it in its little knot of brain in the form of STM for some small periods of time.

Note: it is worth noting here that most of these animals (from insects to mammals) can not only perceive the sensations applicable to human beings (such as light, sound, touch etc., as humans do), but their brains are equipped with special molecular gadgets to sense certain extraordinary sensations, such as geomagnetic fields, ultrasonic sounds, infrared rays, etc. (see Ch. 4, p. 85) – these faculties give them their immense ability to hunt for their prey in darkness or to migrate to distant lands. Humans can remember things such as faces or facts for a long time; likewise, these animals can remember geomagnetic fields for a long time.

Many insects have better perception than worms, and can remember their external world for a slightly more extended period of time, as evidenced by the way they perceive and remember their nests and sources of food. All vertebrates, such as fish, reptiles and birds, show a remarkable degree of perception and memory, which means that they have some improved degree of memory to carry out their survival mechanisms. As an example, it is common to observe pigeons to come for their daily grains to a particular place; or a falcon has to remember the course of its flight to return (using 'geomagnetic memory'). All mammals have excellent short-term and working memories. A mouse has to possess a complicated working memory and some LTM to move around in a maze. All advanced mammals have some great degree of LTM, and LTM improves as they ascend in phylogeny (e.g. elephants, apes). Consequently, it can be said that the complexity of the CE in animals has improved progressively to enable them to perform more complex actions.

Finally, in the case of human beings, the complexity of memory and the CE has reached its zenith and conferred on humans their intellectual abilities.

Free Will

Memory, volition and the CE are so well developed in man as to make his actions appear to be totally free of the will to survive. We will see below that the will to survive is pushed to the back seat at times in humans, and volition is almost completely freed from will – and this is the correct meaning of the term *free will* (see Ch. 10). It would appear that there is much free will in most of our actions and we can consciously make choices. However, it must be realized that the basic will to survive, which lies dormant deep inside us, tirelessly works in the background at all times, only to surface when there is danger to life, as shown earlier.

WILL *VERSUS* VOLITION

Hitherto, we have seen how the will to survive has progressed into the development of volition, memory and the CE in advanced mammals. However, in man (and also perhaps in some advanced animals) the well-developed volition and the deep-seated will compete with each other to play a dominant role in managing challenging situations, and this results in certain problems, as shown below. This is the problem of will *versus* volition.

The Conflict between Will and Volition: The Origin of Emotions

In animals the volition generally 'agrees' with the will, and the will always does what is best for the animal's survival – a tiger cub in a jungle may play around with its brethren cubs (an act of volition) but they constantly divert from their play to look out for danger (an act of will to survive, whereas a human child may play quite oblivious of its surroundings; see below). These animals may live in groups but with a constant competition for food, shelter and mate – they look out for their own advantage but not for their collective good; for example, it is a common observation that a herd of bison will not fight collectively to overpower a tiger's attack, but they simply let the tiger kill one individual bison and appear quite unconcerned about it. In higher animals, such as primates, volition may display all sorts of mischievous acts, but even these animals are never far removed from the will to survive. It can be observed that they periodically and constantly revert back to expressions of fear and rage during play – fear, rage etc. are the expressions of survival.

The behavior in human beings (and perhaps in some advanced primates) is different; for example, a person can feel rather more secure in a large group of people than feel a threat to his/her survival – they tend to demonstrate civilized behavior. Humans are actually helpful to each other even at the cost of their own discomfort, and sometimes they go to the extreme of becoming altruistic, ignoring their personal benefit. A human being can indulge himself/herself in complex passions, such as play, pity, compassion, charity, sacrifice, jealousy, cheating, avarice, etc., and become devoted to these passions. What a wonderful display of volition humans have! And at the cost of their own will to survive!

However, it must be noted that deep inside each human being there is a strong will to survive, but it is not allowed to be overtly expressed. This will to survive is suppressed constantly by an individual's powerful volition – this enables humans to display all the above complex passions. Nevertheless, the will persistently fights back to express itself, which results in a conflict between the two. For example, an employee wants (or '*wills*') to say something very frankly to his/her boss, but rather keeps it 'under his hat' because the conscious rationale *voluntarily* tries to avoid a battle with the boss, lest it should cost his/her job! A boy's *will* may want to keep the candy for himself, but his *volition* may want him to be generous, as good society prescribes. These conflicts of interests result in anxiety. An another example is the instance of a father who *wills* to avenge the murder of his only son by slaying his murderer, but he *stops* himself just short of committing this murder because of the fear of punishment and its consequences (rather he seeks justice in courts of law) – these suppressed desires take their toll. However, occasionally, when a person's passions are roused to heights (passions are volitions!), the prudent restraints of society become too difficult to endure – a father's passion to avenge the killers of his son sometimes surpasses all social etiquette and commits him to follow his wilful action.

The moral beliefs we possess deep inside us have always conflicted with the will (see Ch. 10, p. 197). This is to show that voluntary restraint of action goes against the will, resulting in discrepancy. The will to survive always tries to abolish the act of volition by remorselessly cautioning us of only what is good for the individual self – not caring for a collective good. However, in human beings the well-developed volition perpetually nags our wills to do this and that – like children who incessantly pester their mothers to buy them an attractive cotton candy at the fair. In short:

> Volition proposes; will disposes.
> Or, will proposes; volition disposes.

What is the end-result of this discrepancy between will and volition? These are irksome self-cancelling vacillations, and these conflicts forms the basis of emotions which are seen to affect humans so commonly and so devastatingly.

Psychosomatic Disorders

Most of the negative feelings we experience (anger, jealousy, hatred, etc.) are the result of a conflict between will and volition – this is perhaps the price we pay for possessing an advanced CE. On the other hand, pleasant feelings happen to us when our volition concurs with will – in other words, we feel happy when we get what we want – and become angry when we don't. However, there is some evidence that advanced mammals such as elephants and primates also experience some of these ill-feelings, which is hardly surprising, owing to the fact that their volition is also very well developed and tries to act independent of their will (as discussed in 'The origin of consciousness and mind', above), whereas, obviously, an earthworm or a toad cannot have any ill-feelings because it has only will to survive.

Over a period of time this *psychological* conflict results in *physiological* changes in the somatic (bodily) functions (hence the term *psychosomatic*). When the will is suppressed by voluntary acts, it responds by causing bodily changes to cope with the stress. We saw in Part I (Chs 1–3) of this book that the thinking cortex has a great influence on the hypothalamus, which in turn affects the body through the HPA axis (Ch. 1, p. 23), which results in a hormonal imbalance (see Papez circuit in Ch. 3). Thus, when you negate with your boss your hypothalamus injects adrenaline into your system, causing your blood pressure to rise a little, and over a period of time this leads to systemic hypertension. A heavy-duty truck driver who is in a constant state of jitters puts his autonomic nervous system under stress and this affects his enteric nervous system (Ch. 1, p. 5), causing ulcers in the stomach, intestinal motility problems, etc. Therefore, these conflicts appears to be the fundamental cause of various psychosomatic disorders, such as ***hypertension, irritable bowel syndrome*** (IBS), ***peptic ulcers, anxiety disorders*** and many more, which are observed so frequently nowadays. The cure for this may be found in harnessing the volition and subjugating it; perhaps this is the basis for the healing power of yoga and meditation.

When the Will Loses the Battle!

We have seen that the will to survive has an unimpeachable authority over the life of an individual organism – it tries to sustain life against all odds. However, in the case of humans it appears that, in some special situations, the will to survive is given second place, usually in favor of an altruistic act for the common social good or for the benefit of a fellow human being (see below). How is it possible that the will has taken a back seat in these circumstances? Below are a few common situations where it appears that the deep-seated will is conquered.

Suicide

It is strange that human beings can willfully take their own lives. What happened to the will to survive that seems to persist against all odds in an organism? How was it tricked to enact self-destruction? However, it appears that in most cases of suicide the will to survive is revived at the last moment (alas a little too late), so that life makes one last attempt to reverse the killing process. For example, in a case of suicide by drowning it is a common observation that the individual had tried to grasp at anything before the event of death. According to a story, which is certainly apocryphal, Diogenes the cynic (*c.* 400 BC) had put an end to his life by holding his breath – we now know that it is impossible to commit suicide this way, because the subconsciously present will to survive overrides the voluntary action of holding the breath so easily that it is impossible to kill oneself this way.

Sacrifice in Passion and Altruism

Another sort of self-sacrifice is that which is done in passion. We occasionally hear about the sacrifice of a mother who rescues her child from drowning but accidentally drowns herself in the process. A soldier sacrifices his/her *self* for the benefit of his/her nation – no doubt because of the pitiless military training they have experienced. In the same way, fanatical sacrifice of one's life for religion or ideology is also surprising. But these are only sporadic incidents of the victory of volition over the will to survive. Perhaps in most of these cases, the underlying *biological meaning* of the sacrifice is an attempt to preserve their own race or clan at large, and this instinct can be understandably stronger than individual survival.

THE WINE-GLASS ANALOGY!

In the end, we can say that our CE directs our sanity. It can be stated that when primary thoughts and ideas are arranged in an orderly fashion it constitutes a *sane mind*, and when they become disorderly the conscious mind becomes *less sane*. And when the order becomes intolerably upset, *insanity* results. We can playfully compare the central executive in a human being to a delicate wine-glass half filled with wine:

Optimists say the glass is *half full*
Pessimists say it is *half empty*
The wine glass splits into two in *schizophrenia*
It gets shattered into multiple pieces in *multiple personality disorder*
It gains a glint in *'superior-than-thou' personalities*
It loses its sheen in *depression*
And now you can make *your own cocktail.*

Now to Chapter 10, which discusses some philosophical aspects of thought and shows how the molecular-grid model can help to unravel the age-old problem of mind-and-matter. In short, it will treat ***thought as a metaphysical phenomenon.***

CHAPTER

10

Metaphysics of Thought

OVERVIEW

Any work on thought is incomplete unless we touch upon the essentials of the problem of mind and matter, hence this chapter. The question of the nature of thought has intrigued man for centuries. Throughout the Middle Ages the study of thought or mind was the prerogative of religion, whose indelible marks are obvious even today. In this modern age, science has made tremendous progress and has managed to bring some of the abstract fields of the study of mind, like that of psychology, into the scientific realm. However, the actual study of the nature of thought itself still lies in the domain of philosophy, which has an important quasi-scientific branch called ***metaphysics***, and this chapter encompasses a discussion on metaphysics and its main divisions. The ground of metaphysics is rather slippery to tread on, and this chapter is only an introduction and offers something of a hand-rail for the reader to catch up with its general idea. We will also study the problem of 'mind-and-matter' at the end, and see how the molecular-grid model fits into this discussion.

We have seen that the human mind is vastly dependent on the sensations it receives from the external world (Ch. 9. p. 169). Consequently, for a thought to form, a stimulus from the external world is essential (see 'Note' below). The mind receives the external stimuli and converts them into internal perceptions – the mind is 'internal' (the ***observer***) and the surrounding nature is 'external' (the ***observed***). Thus, there are two obvious aspects for studying human thought – the observer and the observed. The observer is the human mind, and the observed is the surrounding nature. This chapter discusses first the enigma of the external world, then the nature of the human mind and, finally, the relationship between the two, i.e., the paradox of mind-and-matter. The relevance of the molecular-grid model to the mind-and-matter problem becomes apparent at the end.

Note: There are two fundamental divisions in the study of thought in philosophy – *empiricism* and *rationalism*. Empiricists argue that human thought is greatly dependent on the sensations received from the external

The Biology of Thought.
DOI: http://dx.doi.org/10.1016/B978-0-12-800900-0.00010-5

world. The preceding chapters have taken this view and this has suited our scientific pursuit of thought very well. However, the second division encompasses the intriguing concept that the human mind is much more than a sensory experience. For example, a person observes a ball lying still on a table – consider that the table has started tilting, and this causes the person to anticipate that the ball will roll down and fall, even before the ball has actually started rolling – whereas a dog by his side will not anticipate this. This means that humans are capable of extrapolating future events from the known facts. The intellectual achievements of man are considered as examples of rationalism. We will not go into the details of this, but continue with the empiricist view in our scientific pursuit.

THE PROBLEM OF THE OBSERVED: THE ENIGMA OF THE EXTERNAL WORLD

The external world constitutes everything that we observe in nature, everything that is external to our sense of perception; for example, when a person looks at a sparrow perching on a window ledge, the *person* is the *observer*, and the sparrow, the window and the ledge and everything else constitutes the *observed*. These external events may be studied in two possible ways – the events of the external world may be ***deterministic***, i.e., they may go on and on without any relevance to the intervention of human agency – or they may be quite ***indeterministic***, i.e., the events may change in accordance with the actions of human beings or other agents. These concepts are discussed below.

Determinism

A Traveler's Dilemma

A traveler who was walking through a forest has a choice of taking one of the two diverging paths available to reach the same destination (see 'Free will: yes or no', below, for a discussion on 'choice'). As the traveler started walking on path-1 a thorn pricked her foot and, thinking that the first path might be littered with several such thorns, she changed her mind and took path-2, and as she was treading along the second path a mad elephant trampled upon her and broke her leg. Now a question: what was the *real* cause of her broken leg? Was it the elephant? Had the traveler pursued path-1 she would not have had the misfortune of encountering the elephant. Or was the thorn really responsible for her misery, which had pricked her on the first path and caused her to take the other path? Whatever caused the thorn to be there on path-1 must be the *real* cause of the traveler's broken leg. If the thorn had not been there, she would have continued in the first path (in which event, of course, we cannot predict the other possible outcomes, see 'Indeterminism', below). Surely then, whatever caused the thorn to be there can be presumed to be the real cause of her broken leg. As it happened, the thorn was lying there because a farmer working in a field by the side of path-1 had noticed that a thorn had pricked his little daughter's finger as she tried to pick a rose, so he removed the thorn and nonchalantly tossed it away, and it so happened that it had fallen on to path-1. Now think of this: shall we blame the father for his rash act of throwing the thorn on to a footpath, which ultimately led to the traveler's broken leg; or shall we hold the farmer's little daughter responsible for her negligent act of picking the rose, which if she had not done would not have led to the traveler's broken leg? Or shall we blame the rose itself for being there in that place in the field? So on and so forth. We can carry on the speculation backwards until we reach cosmological angles!

In a similar manner, how can we ascertain the *actual* reason for the joy of a carefree boy running across a playground, what caused him to be so happy? Or the real reason for the sorrow of a lugubrious old woman sitting under a tree? A deep analysis of a jovial event, on one hand, robs the pleasure we can derive from that event – philosophy can be so disgusting. However, on the other hand, such deep understanding of an unhappy event also mitigates the pain that it may entail. The following examples throw more light on determinism.

Zeno the Stoic

Zeno the stoic (334–262 BC) was once beating his slave for some trivial offence and the slave protested, saying that by his master's own philosophy he was destined to commit the folly and pleaded excuse because it was not actually his fault; to which Zeno replied, with the calmness of a sage, that on the same account he was also destined to punish him for it (Durant, 2006, p. 125). Look at that, how can we escape from the inexorable clutches of this deterministic world!

Cleopatra's Nose

Mark Antony (83–30 BC), the great Roman military commander, friend of Julius Caesar and lover of Cleopatra, the beautiful queen of ancient Egypt, was on his way to the decisive battle of Actium. It is but an irony of fate that, before setting out for the battle, he happened to stare at a statue of Cleopatra and became transfixed by her beauty, and lingered on, amazed at the sculptor's perfection, which diverted his mind and caused him to use irrational battle tactics and thus lose the battle grievously. The defeat at the battle of Actium led to the rise of the first Roman Emperor, and this is considered crucial for the concatenation of events that followed and shaped the history of Europe, and consequently that of the world. In his *Pensées*, Blaise Pascal (1623–1662) remarked, 'Cleopatra's nose, had it been shorter, the whole face of the world would have changed.' Can we say that the fate of the world had hung in balance on Cleopatra's nose? Or shall we say that the present history of the world was decided by that bygone sculptor's hammer and chisel!

Stephen Hawking

Professor Hawking said in one of his lectures, 'Does God play dice', that:

> The idea that the state of the universe at one time determines the state at all other times, has been a central tenet of science, ever since Laplace's time. It implies that we can predict the future, in principle at least... A butterfly flapping its wings can cause rain in Central Park, New York. The trouble is it is not repeatable...

Think of this, after all, it is such an interminably interconnected world!

The contention of determinism, finally, is that all events are determined and nothing could have changed. One event precedes (or follows) another event in an inevitable chain of events in the external world. Determinism can be extrapolated into the future, in which future events can be predicted, because it is a die-cast situation and the future events *must be* as determined as the past events.

A Word of Caution!

We are also struck by the everyday occurrences of such phenomena and this leads us to think that, 'it could have been like this, if it had been like that' – but no amount of musing leads us anywhere. To discuss determinism exhaustively and to an extreme end is pointless, as it takes us nowhere and serves no purpose. As is also the case with the uselessness of predicting the future. After all, determinism cannot be equated to *predestination*, which has a strong religious connotation. Determinism, followed to its extremes, soon leads us into the murky labyrinths of its cynical aspects, such as *fatalism* and *defeatism* (**pessimism**), which may soon presage undesirable philosophical indifference and inaction (***stoicism***). Nevertheless, the essence of determinism has to be understood to grapple with the issues of *epistemology* (see 'Practical implications', below).

Indeterminism

Alternatively, these external world events can be explained by considering that all these occurrences are mere chance events and could have happened in many other possible ways, and our traveler in the tale was really lucky that she did not pursue path-1 intrepidly, in which case she might have been eaten alive by a hungry tiger waiting on that path. This is indeterminism, meaning that the external world is not determined, and can happen in any number of possible ways. However, it is doubtful that throughout the ages science could have depended on indeterminism to conduct its inquiry of the external world – the world becomes too chancy and uncertain out there!

Having discussed the external world and its nature, we shall proceed to examine the nature of the observer himself.

THE PROBLEM OF THE OBSERVER: THE PARADOX OF THE HUMAN MIND

Before going into these aspects, we must first settle upon the issue of reliability of our own senses – how far can we depend upon our senses in judging the external world events, as shown below.

Realism *versus* Idealism

Philosophers have long been divided on the actual nature of the external world and the ability of humans to detect changes in the outside world using their senses (look at the example below to see what this really means). ***Idealism*** is the doctrine that a person's thoughts alone are the fundamental reality and the external physical world is only a reflection of our ideas, and there is no way in which we can know the *real* nature of things (Durant, 2006, p. 354). Extreme versions of idealism even deny the actual existence of the external world outside the observer's minds. In contrast, ***realism*** is the doctrine that the external world exists independently of us; that is, the events of the external world go on and on, irrespective of the observer's status (Durant, 2006, p. 695). This is explained below with the help of an example.

The Case of a Taxicab

The nature of the external world is known to us only through our perceptions – that means we depend entirely upon our senses to know the exact nature of the things we observe. Only when we open our senses to the external world do we know that there are changes occurring outside us. When we close down our senses there is no way in which we can say whether certain happenings are really occurring outside us.

For example, when a taxicab passes close by your side, you are actually *visualizing* the cab, *hearing* its buzzing sound and *feeling* the sudden gust of wind – all these things make you believe that a cab has passed by your side. Close down all your senses (the eyes, the ears and the general tactile senses), and the result is that the cab has disappeared, the booming noise has ceased and its movement has stopped! Now, can you say for *sure*, with any authority, that a cab has passed by you? The only evidence that the cab had passed is that it might have shifted its relative position from one place to another 'a moment later', and that can only be known by opening your senses again 'a moment later'. But right now, *at this given moment, when you had shut down your senses,* there is no way of telling this way or that way whether the cab existed or not.

Now, the point is that these external world events may either keep happening irrespective of the observer's status (realism), or the events are only our mental perceptions and nothing more (idealism). The ***realists*** argue that the cab has *really* gone by at this moment, only that we could not have known, because we had shut down our senses. The ***idealists*** argue that the cab might or might not have passed by at that particular moment, and there is no way we can be certain, and hence it does not affect us in any way. 'Pure' idealists may even claim that, at that moment, the cab itself might not have existed from the observer's standpoint.

Practical Implications

Of course, when we observe such things as cabs moving in our day-to-day life there is no confusion with our perceptions, as we have *some* 'ingrained' experience with these 'common-sense' deductions. What happens is that we observe *something* and we conveniently *presume* the rest of the events. Even if we have closed our eyes, if the other two senses (in the above example) are working, we can deduce that a cab has passed by using our past experience.

The problem arises when we have to examine things or facts that are very new to us, or things which are very subtle (such as the study of the human mind itself), or are very far away from us (for instance, at astronomical distances). These experiences have no set standards to compare – our intuition does not work here. In such circumstances there comes this antithetic difficulty in interpretation. These subtle differences are all the more important when we start seeking issues such as the ultimate truth of matter (***ontology***), or the ultimate truth of mind, or the relationship between the mind and matter (***epistemology***).

How Dependable are Our Senses?

How can we confidently rely upon our senses and our experiences when we perceive the external world in a different light from the actual fact? Take, for example, the case of a motion picture – the undulatory movements of our favorite heroine are in fact a serial of individual *static* frames, which when moved with certain rapidity (say, 24 frames per second) can trick our senses into seeing a beautiful dance – mankind is so adapted to this trickery that a colossal film industry has flourished down the ages. To give some more examples, take the illusion created by mirages, the incidence of parallax error of an observed image or various optical illusions enacted by magicians, etc. How far can we rely either upon the human mind or upon its judgment of the external world to know the *real nature of matter*, the so-called *thing-in-itself* (or the *noumenon*)? It has also been shown that there are differences in individual perception – each individual perceives an object or event differently, and thus our experiences themselves are not totally uniform because of individual genetic variations (see Ch. 4, pp. 80, 82 & 90 and Ch. 7, p. 140).

Is Realism True?

Perhaps there appears to be some grain of truth in realism. Consider this: if two taxicabs collide, not very far away from us, we cannot call the event to be only our mind's perception; neither we can say that matter in the form of those cabs does not really exist except in our mind's 'eye'. In reality this event affects not only us but also our fellow observer (may that be our pet dog), because both of us run away from the impact of collision; and the event also results in some *verifiable* physical change in the external physical world in the form of a crash and its resultant debris. Thus, it can be safely assumed that the external world events go on independently of us – because the whole of these things and the resultant changes are verifiable by each other. It must be noted that the entire human history and its scientific progress depends on this realism.

Or, is it Idealism?

While on a small scale, as in the above example, realism appears to hold some water, on an astronomical scale idealism appears to gain ground. To consider the uncertainty of existence of matter in the external world look at the following analysis. What do we know of this *endless* universe? Imagine this: it is difficult to say how big the universe is because we cannot even *imagine* how big it is. Our Earth is only a tiny, tiny part of the solar system, and our solar system is a tiny part of our Milky Way galaxy, and there are about 170 *billion* such galaxies in this *observable* universe. Where is all this vastness situated? Imagine that scientists of tomorrow were to discover that this whole thing is encased in a big eggshell-like mass (as the ancient Chinese had believed), and we are again at liberty to ask 'where is this eggshell located?' This question/answer continuum is seemingly endless – this is ***infinity***. If you give a convincing answer that the universe is *expanding* and there is no real outer boundary, then it automatically means that, in that case, there *must* have been a *definite* beginning in the past (starting point), and this is the reason for the present state of expansion. If there was such a beginning (the ***Big Bang***) then it is a *finite* entity, and thus there must have been a *past*, but the events before the Big Bang are entirely unknowable or uncertain. If the Big Bang has really occurred, what caused it, or what was the *First Cause*? Hence, for the universe to be real, it *must* be finite, because infinity is *unthinkable* to the human mind. But, at the same time, the *'reality could not be really real'* because we cannot find an end to the universe.

These insoluble dilemmas are called ***antinomies***, which bespeak the innocence and immaturity of human knowledge, apart from the hard limitations of the human mind's design in these matters of perception and knowledge. Why is that the human mind cannot think of these infinities? Why are we so uncomfortable with these half-truths? Immanuel Kant, in his famous book, *The Critique of Pure Reason* (1781), came up with an answer: he maintained that these dilemmas arise because we tend to suppose that space, time and cause are purely external and independent of our perception (Durant, 2006, p. 344–357). Instead, he integrated space and time with human perception and asserted the primacy of human thought in the questions of mind and matter. The whole of the universe, he said, is represented in 'you' (*the transcendental idealism*). In other words, 'you' and the universe are one and the same, or meaning 'you' are 'the one and the only reality' (***solipsism*** = *'you'* is *'solo'*). (Thus, Kant has approached *'consc-ious'* human thought with his *'Kant-ious'* idealism.) What an egocentric philosophy! However, the undeniable evidence that the human mind draws heavily on primary thoughts gained from experience (pp. 89 & 169) shows the reality in no uncertain manner – that nature created us, and not *vice versa*.

The ancient Hindu philosophy also describes a metaphysical doctrine called ***Advaita Vedanta*** (as expounded by Adi Sankaracharya, *c.*788 AD), which deeply dwells on the mind-matter duality and is a representative of idealism. In essence, Advaita is a non-dualistic philosophy which shows that the knower and the known are the same, and has asserted that the *Brahma* (the Absolute Truth, the knower and known together) is *Sat-chit-ananda* (Pure Existence – Pure Knowledge – Pure Bliss).

Having examined the limitation of human faculties in deciphering sensations, we continue to examine the intricacies of the human mind.

The Paradox of Free Will

Free will is the capability of a rational agent (human being) to exercise freedom in acts of conscious choice from the determining compulsions of the human body, environment and circumstance (Durant, 2006, p. 694). Free will is the power of acting without the constraint of necessity or fate. The exercise of free will is thought to be a prerogative of the rational human mind, and the egotistic nature of the human mind always presumes that human choices or actions are not determined and that they are capable of dictating their own destiny. The development of free will is discussed in Chapter 9. But the very existence of free will has been a subject of debate right from Aristotle to the thinkers of the present day. René Descartes (1596–1650) contended that the human will is so free that it can never be restrained. On the other hand, philosophers like Spinoza and Schopenhauer believed that the will cannot be free at all. The confusing concept of free will is explained with the following example.

Free Will: Yes or No

One day, my friend, Raja, attended a banquet dinner where a variety of exotic dishes were served. Naturally, his keen sense of appetite was aroused and he would have loved to tuck in to all the rich food. But, at the same time, he was worried about the extra calories adding to his already great paunch and finally thought better of it, and avoided eating the stuff. Now, a metaphysical question: what *really* caused Raja to *avoid* the banquet? On one hand, he *wanted* to feast at the banquet, and on the other hand he did *not want* to feast for the sake of his health. Examining Raja's thoughts closely, *two possibilities* can be considered that made up his mind to arrive at the decision of avoiding the banquet:

- ***No free will.*** It was possible that he did *not really want* to eat this elaborate meal – his will had already dictated his thoughts to avoid the meal (= no free will). His deep seated will stood strong, negating volition (see Ch. 9, p. 176).

In this first instance, his will was fixed and had no choice, and was the ***slave of his constitution*** or the ***chemistry of his mind*** (he couldn't have wanted to eat the banquet at all in the first place); in other words, he had no choice but to avoid the starchy meal – the idea of eating the rich food was only a passing whim, never a serious consideration

- ***Free will present***: It was also possible that he *really* wanted to eat it but did not eat it for his health's sake – his *free mind* could *'bend'* his will (= free will present). His volition stood strong, negating the deep seated will (see Ch. 9, p. 181).

In this second instance, Raja had a free mind and he was free to choose, and he was the ***master of his thoughts***. He wanted to eat the food, but better thoughts had prevailed and made him to *choose* to avoid eating the stuff. His volition (free mind) had taken over and bent his deep-seated will.

In the case of our traveler in the example given above, she seemingly has a will of her own to make a *choice* of selecting one path from the other, but in fact, such decisions are based on an individual's past experiences, and are also heavily drawn from one's own personal inclinations, so they are ultimately deterministic, or so they appear.

The Illusion of Free Will (No Free Will)

The absence or presence of free will has deeper connotations than the above example (Durant 2006, p. 226–230) – the absence of free will (that is, the will is fixed) *dictates* that a person is *destined* to have an *incorrigible* and *uncorrectable* personality. As an example, some persons *by nature* are bold or timid, gentle or cruel, truthful or delinquent – that a person is *constitutionally* bound to behave that way and he cannot be sincerely held either in praise or disdain for being so. One is not morally responsible for one's actions. A preacher of morality may be moral in his actions only because he *cannot* simply be otherwise! How uncharitable is this idea, really!! However, it goes hand-in-hand with determinism.

The Damsel and the Mouse

Once upon a time, a pussy cat earnestly pleaded with a passing sage to turn her into a beautiful damsel, so that she could marry a prince and live happily ever after, and her wish was granted – everything went well for the damsel for some time, until a mouse ran across in front of her, causing the damsel to pounce upon it. Such is also the unyielding or uncorrectable nature of human will, in general.

Altruism Has No Biological Foundations

It must be noted here that you *cannot* wish for anything that does *not really* appease you. If you think you have done a good job by helping the poor with alms, it is only because that gesture has appeased your 'desire to help' – the rich *needed* the poor only to show their generosity! When you help a friend in need, you are really pleasing your own self more than fulfilling his/her needs. A child shares her bar of ice cream with her pal only because it satisfies her will by filling her with a sense of pride.

You May also Consider This:

The hardest trial of the human heart is to test whether it can really bear a rival's failure without triumph! And nothing is more engaging than the topic of another's failures. In the depths of his heart, a person *cannot* act in any way *truly disadvantageous* to them. Your will always caters to satisfy your own biology – everything else is just ostentation. *Altruism has no biological foundations.*

It's Worth Pondering Over This:

We do not like anything because it is good: it appears good to us because we like it. Thus, arguments and debates to avoid wars that were orchestrated by various historically important men across the tables throughout the ages have been examples of useless farce – by nature, men had to make wars – and they did! And it is also true that we tend to listen more readily to the passions of the heart than the logic of the head – and it is a common observation that curiosity overrides caution – people naturally tend to go to a place that is forbidden.

A Word of Caution!

The idea of determinism and the absence of a free will may erroneously seem to absolve the crimes of the hardest felons, holding that they were not strictly morally responsible for their deeds, and so can be exonerated. However, this does not mean that the punishment meted out to the hard criminals should be passed on mildly. The chief function of a civil society is to moderate the wills of their public, in general. The individual idiosyncrasies have to be 'ironed out' by some really persecutory laws to run any civil administration. In instituting suitable punishment to such criminals, the reader may take solace from our ancient philosopher Zeno the stoic (see p. 191).

Experimental Evidence

Some neuroscientists and cognitive psychologists have conducted experiments to empirically prove the existence of free will. Benjamin Libet in 2002 conducted some (controversial) experiments and found (to his surprise!) that actions are already under way shortly *before* a person intends to do it. He suggested that we do not (in fact, *cannot*) *consciously initiate* our actions, but we possess the ability to veto (abolish) such unconsciously generated thoughts or actions (for a complete discussion the reader may refer to 'free will' in the *Stanford Encyclopedia of Philosophy;* O'Connor, 2013). These researchers argue that the notion of free will is simply a deeply entrenched illusion. However, all these abstract experiments are hotly but justifiably debated by other researchers.

Man, The Super Animal (Free Will Present)

The presence of free will makes humans '***superior animals***' among the beasts and ***morally responsible agents*** – because with experience or training an individual can *improve* upon their social etiquette. However, it must be noted that in a perfectly deterministic world, the existence of free will is problematic to fit in, if not impossible. People have tried to accommodate free will with determinism (and called it *compatibilism*); maybe at the cost of considerable effort, but these two things naturally tend to fall apart.

Free Will in Animals

The existence of free will in animals is not considered because their behavior is considered to be often only goal-oriented and does not have moral implications. Nor do they have the ability to choose between alternatives, nor think of their long-term consequences. Their will is much more basic (Ch. 9, p. 175).

HOW IS THOUGHT INITIATED?

The problem of the existence of free will gives us another tangle of questions. How are thoughts initiated in the brain? What *causes* us to think? Are they generated *per se* in the brain, or are they induced from external influences? It was shown in Chapter 8 that the dendritic pleats when excited in a serial manner generate primary thoughts and ideas. It was also shown that the dendritic pleats may stay in place for variable periods of time, causing us to remember something or an event for variable periods. It has been shown that we have memory traces for all memories stored in the brain in the form of dendritic pleats. It can now be presumed that these past ideas may be re-initiated by simple excitation of these dendritic pleats (see Figs 8.3 & 8.4). But again, the big question is what actually *initiates* the excitation? Can the dendritic pleats be excited all by themselves, or do they need an external initiating stimulus?

Let us run another thought experiment to aid in the discussion. Imagine that David was passing an idle hour on a Sunday afternoon, and had *suddenly remembered* his old college friend, whom he has neither heard of nor met for a long time. Now let us put this question. Why did the thought of this old friend enter David's mind *at this moment*? What caused David to think of this long-forgotten friend suddenly? Consider the following possibilities:

A *de novo* Thought

David has every reason to conceive the notion that he has thought of his friend *entirely by himself*, and that this memory of his friend was really *born* in his mind. It was a ***de novo thought*** – originated within him. However, we are at liberty to ask this question: can a thought really arise without any inducement? Can a thought 'take birth' in the mind by itself, without a stimulus from an external input? However, even if such a free thought exists it must be strictly under control of the central executive – otherwise the mind may start all new ideas randomly, which makes the formation of a *sane mind* impossible. The existence of *de novo* thought has something to do with the presence of free will.

Externally-Induced Thoughts

Alternatively, it is also possible that some trivial thing had induced David to think of this friend, such as some other related thought or associated event. For example, consider that this friend of his was an expert swimmer and had always won intercollegiate swimming competitions of their time. On this particular Sunday, at breakfast, David's son had remarked about his forthcoming swimming competition. These remarks related to swimming by his son might have triggered the memory of his old swimming-expert friend – the remarks acted as *external agents* which induced the bygone thoughts into his subconscious mind on this afternoon. Or, it is also possible that our man had seen the picture of a swimmer in a news item in *The Times*, which induced the thought of his friend in his subconscious mind. But David is not aware of any of these external inducements. Thus, this gives an illusion of a thought being born within him (*de novo* thought), but all the time it was only an idea derived from past experience, stored up in his memory, induced by an external agent.

A Third Type of Thought Inducement

It is interesting to know of a third type of thought inducement. These are the thoughts related to internal sensory inputs. The best examples are ***hunger*** and ***thirst***. For example, low levels of glucose and water in the blood induce thoughts of hunger and thirst. The *sexual urge, urge for competition and excellence* appear more complicated and use both external and internal inducements, and *hence these are more conflicting in nature* (see Ch. 9, p. 182). Genes have a potential role in these internal inducements. However, it can be said that these internal inducements are in fact not strictly internal but external, because the observer is the brain (see below), and the body itself can be considered external. This means that we are now going into the perplexing 'mind-body' problem, and we will be dealing with this below.

WHO AM 'I'?

So far we have discussed that 'the observed' is the external world and 'the observer' is 'you'. Let us try to find out what exactly constitutes this internal observer. When you look at a sparrow perching on the window ledge, the sparrow, the ledge and the other details constitute the external world and 'you' are the observer. Obviously an observer is the 'knower'. Now think of this: who *exactly* is the observer? The knower is your brain. The observer perceives things through his sense organs, from where a bundle of sensations are carried to specific areas of his brain; for example, the eye transmits the image of the sparrow to the brain, where it is perceived as an idea of a sparrow (see Ch. 4, 'The concept of primary thought'). Thus, the brain is the actual observer. All the sense organs, such as eyes, ears, touch/pain receptors in the skin, etc., only transmit information to the brain but they themselves do not perceive anything (Ch. 5, 'Pathways of perception') and thus they represent the external world. This can be translated into the following statement:

> The sense organs belong to the external world
> The brain alone constitutes the internal observer.

Thus, the human brain constitutes the internal observer, and the rest of the body (including the sense organs) is external. Now we will dig deeper into the problem of mind-and-matter which has intrigued sages, seers and scientists down the ages. We will try to analyze the following question: how is matter related to mind? We saw in Chapter 9 that the human mind is constituted by streams of primary thoughts. Hence, for the sake of discussion, let us re-phrase the question of mind-and-matter as a ***thought-and-matter*** problem. In other words, we will now see what exactly is a primary thought and how it is related to matter.

Neuron as the Observer

Extending the above discussion a little further, it can be deduced that the actual observer is a particular group of neurons (situated in a specific area of the brain – such as visual neurons in the visual cortex, or olfactory neurons in the olfactory cortex, or general sensory neurons in the somato-sensory cortex, etc.), and finally, it can be said that ***an individual neuron*** (with all its dendritic processes and synapses) itself represents an observer (in miniature form). In the above example, each visual neuron perceives a tiny bit of the total sensory input from the sparrow, the ledge etc., and this visual neuron is the internal observer for visual signals. Similarly, if you are listening to a piece of music the auditory neurons perceive the tiniest part of the music and auditory neurons are the observers for sound signals; and for smell sensation the olfactory neurons are the observers, and so on. Thus, for all sensations the corresponding *perceptual neurons* (see Ch. 5, p. 106) constitute the internal observers. Hence:

> A perceptual neuron is the internal observer.

Which Part of Neuron is the Real Observer?

The 'whole neuron concept' has already been shown to be no longer tenable (see Ch. 6, 'Where in the neuron'), and it was also demonstrated in Chapter 6 that the cytoplasm and nucleus do not take part in receiving and deciphering the signals directly, it is only the neuronal membrane that takes part in the conversion of sensation into perception. This can be interpreted in the following way – the cytoplasm and the nucleus belong to the external world and thus they do not constitute the observer. And

Chapter 7 has shown that the molecular grids are situated in the dendritic membrane of a neuron and from here the primary thoughts are generated. Thus, we can state the following:

> Cytoplasm and nucleus are external agents. Neuronal membrane is the internal observer

Membrane as the Observer

We will consider the following observation to further establish the importance of the neuronal membrane as an internal observer. If you supply neurons with certain psychoactive drugs (stimulants such as caffeine, LSD, amphetamine, or depressants such as Valium, alcohol) they modulate the perception of a sensation to either a greater or a lesser degree; for example, LSD induces 'vivid' hallucinations to external stimuli (a single ray of light may be perceived as a terrifying array of lights) and alcohol induces depression (even a search-light shone into the eyes of an inebriated person may fail to elicit a response). This means that these agents (or drugs) can modify the response to a stimulus, therefore these agents can be considered external. These drugs are carried to the neuronal membrane by the cytoplasmic proteins and this re-confirms our view that the cytoplasm is an external agent.

The *genes* which modify a response to a stimulus also act only through chemical mediators mediated by the cytoplasm (see Ch. 7, p. 131), hence the genes can also be considered to be external. However, it must be said that the genes are the 'modifiers' of the internal observer (or 'mind') (see Ch. 4, p. 91 & Ch. 7, p. 140), but they do not represent the internal observer itself.

Furthermore, it is seen that the response to stimulus can also be altered by changing the ionic permeability of the membrane surrounding the molecular grids (Ch. 7, p. 137). As an example, general anesthetics or local anesthetics (e.g. nitrous oxide, the so-called 'laughing gas', or lidocaine) act by directly modifying the membrane's ionic permeability and inducing sleep or causing numbness. In fact, all the psychedelic drugs described above act ultimately by altering the ionic permeability of the membrane receptors of neurons.

What exactly is internal, who is the observer? The only things that can be considered internal are the ***molecular grids*** situated in the membrane and generating primary thoughts. They generate primary thoughts when they are excited by the ***graded potentials*** coming from the external world (see Ch. 7, p. 139). Upon careful consideration we can say that these membrane potentials are generated originally at the peripheral receptors by the external stimuli and are then transmitted to the brain (see Ch. 5, p. 94), and thus it can be said that *membrane potentials represent the sense-experience* – i.e., the external world. Hence, we may say that we sense the external world by way of membrane potentials. Thus, the following statement can be made:

Molecular grids represent the knower (the **observer**)
Graded potentials represent the external agents (the **observed**).

The Ultimate Paradox

We have concluded empirically that the neuronal molecular grids constitute the observer or the knower. But even the molecular grids can be considered external to the observer, if you ponder over the following observation. It is shown clearly that the characteristics of a sense-experience are essentially dependent on the composition of proteins and lipids in a molecular grid (Ch. 8, p. 160) – each sense-experience has a specific type of molecular grid. This means that the molecular grid only assists the membrane potential to transform into a specific primary thought, and the primary thought itself is a 'modified' membrane potential. Thus, the membrane and its molecular grids can also be considered external!

Now we can consider the ***metaphysical conundrum*** of what is mind and what is matter – and how they are related. Let us rephrase the above discussion into another statement:

Primary thoughts are the modified membrane potentials.
OR, Primary thought and membrane potential are one and the same!
OR, The external world and the internal world are one and the same!!
What was it, again?!

That is, *all these things are external and influence your thought or mind* or 'you'. So what is the internal observer? Nothing material!? The internal observer is only 'pure energy' or 'pure thought', the rest of the body of the neuron is external. What, then, is pure thought? The neurons and their membranes are only external agents through which the internal observer (or the mind, or the primary thought) operates. If we are keen enough about this epistemic uncertainty we can ask the following question: what is the 'cause of this mind', and what is the 'effect it shows on matter'? Is ionic transfer across membrane (i.e. the generation of membrane potential) the cause or the effect? What is mind, and what is matter? Does mind exist without matter? Or is the pun 'No matter never mind' true? Is the knower matter at all? Who then is 'you'?

It is a great pleasure to summarize the whole of this metaphysical problem in the words of the great master himself – René Descartes – who answered this epistemic question of '*Who am I*' in a nutshell:

Cogito ergo sum – I think, therefore I am.

A Final Note: Roads to Quantum Consciousness

It is the author's hope that the above discussion paves the way for a scientific study of thought and forms a suitable basis for experimental biologists to work on the biology of thought – and to elevate research on mind and consciousness to the quantum level. Membrane is the place where these researchers need to team up to decode the secrets of thought, and where the scientists have to apply the known laws of quantum physics to unravel the mystery of mind. They should also realize that quantum interactions at the ion channels and lipid rafts hold the key to many answers. The answers to most of the metaphysical questions lie in the exact place where membrane potentials are created, i.e., in the ion channels, and it is our enduring task to know how electrical energy of ionic disturbances transform into some unknown energy of primary thoughts. Finally, the answer to the question of *why* they are created in the first place is perhaps purely spiritual, and appears transcendental at best.

P A R T IV

THE COMPUTER AND THE BRAIN

CHAPTER

11

Molecular Grids and Computers

OVERVIEW

This chapter examines some of the important similarities and differences between a human-made computer and the brain. The computer is certainly the most innovative product invented by human genius. Computers have revolutionized human civilization and have assisted man in making significant advances in the fields of science, medicine, commerce, literature, humanities, and every other possible field of human civilization. Although most of the time the computer is the slave of humans, sometimes it assumes the role of master. In some fields computers have virtually replaced human intervention and have become bosses themselves – such as the electronic gadgets in *Curiosity*, the robotic spaceship to Mars.

Conventionally, the brain is compared to a computer; however, in this chapter we will see how a ***neuron*** itself can be compared to a computer. First, the basic function of a computer is described, and later a comparison is made between the computer and the brain, especially highlighting the comparison between a molecular grid and a transistor.

THE ESSENTIALS OF A COMPUTER

The 'brain' behind every computer is the ***microprocessor*** (also called ***integrated circuit*** or, colloquially, the '***chip***'). Microprocessors are, in fact, electronic devices that are made up of several tiny ***transistors***. If neurons are the building blocks of the brain, transistors are the building blocks of computers. These transistors are the real electronic microgadgets inside a computer that operate the computer language (the 'software', see below), upon which all modern digital computer transactions are based. A chip may measure anywhere between a few square millimeters to around 350 mm^2, and a modern chip may house billions of transistors!

The Biology of Thought.
DOI: http://dx.doi.org/10.1016/B978-0-12-800900-0.00011-7

Transistors

Transistors are small electronic circuits that control the movement of electrons across them in one direction only, and this is the basic property of any electrical circuit. These tiny electronic circuits are used as signaling units, as shown below. How are transistors made? The wonder lies in the versatile properties of ***semiconductors***, which allow us to 'gate' the flow of electrons through them and thus generate signals that can be made to pass with tremendous rapidity as the current flows.

What are Semiconductors?

Elements such as *germanium* and *silicon* are insulators for electricity (bad conductors), but when they are mixed (or '*doped*') with minute amounts of 'impurities', such as phosphorus, arsenic, boron or gallium, they acquire a peculiar property of partially conducting electricity. Or, in other words, they become *semiconductors*. Semiconductors are of two types, depending on whether they allow the electrons to be added when electricity is passed through them, or allow the electrons to escape from them. In the positive type (***P-type***) the semiconductor absorbs electrons when current passes through it; conversely, in the negative type (***N-type***) the semiconductor gives out electrons to the adjacent field.

Diodes and Triodes

Now, place very thin sheets of N-type and P-type semiconductors side by side, and the result is a ***diode***, which is also a device that allows the current to flow in only one direction. Transistors are created by using *three* layers, rather than two (hence called ***triodes***), and they can be arranged in two sandwich patterns – NPN or PNP. When a small current is applied to the central unit (*the base*) one of the two things may happen – it can either amplify the passage of current across the whole unit, or it can hinder the passage of current. Thus, each electrical wave that passes through the transistor can be represented in two states – ***on*** and ***off***. In other words, it can act like a switch.

The 'Languages'

Digital Systems

Any information in the form of sound, light, temperature, pressure, etc., can be converted into electromagnetic signals, and these signals can be converted back to their respective forms. For example, when we input sound signals into an electronic device, the sound signals are converted into electronic signals in the computer chip, and when these electronic signals are passed to a suitable output device they are again converted into sound signals. Another example: light signals are stored in the chip of your digital camera and when connected to a suitable display mechanism, a picture is again shown. These electromagnetic signals can be studied in two ways – as ***analog signals*** or as ***digital signals***. An example is that of a wall clock, which may indicate time in two ways – a continuous mode in the form of a dial with hands, called analog clock; or in discrete signals of numbers, called digital clock. Modern computers use digital signals for efficient communication.

Binary Language

This language was developed by George Boole, as long ago as 1854. We know that the *decimal system* of counting numbers uses 10 digits – from 0 to 9 and starts from ones, tens, hundreds, thousands and so on; whereas in *Boolean logic* the numbers are counted in a *binary system* (two-digit system), which has only two units – 0 and 1, and all place holders are multiples of two. Any length of number can be represented by using binary language.

Now combine the transistors with binary code. The transistors act as gates to the passage of electricity, switching them to either on or off, and so they can represent the binary language system in two states – ***On*** (or ***1***) and ***Off*** (or ***0***). In other words, the transistors represent an electronic wave that passes through it in two phases, on (1) and off (0), and this feature is used to represent binary code, which also uses only two signals. This means that any input signal (light, sound, etc.) can be converted into a binary signal that can now be studied electrically. Thus, it became finally possible to convert digital language into electrical signals that can be displayed, and this has resulted in the emergence of *electronics*.

Processing Speed and the Chip

Obviously the number of transistors that are present in a device represents the speed with which it can perform. The first microprocessor released by Intel in 1971, the Intel® 4004, contained 2300 transistors on its circuit, whereas Intel's Core™ i7 (the quad) contains over 700 *million* transistors; and when you think of the latest version of chips containing over 2 *billion* transistors, you may imagine what they can do.

As you may realize, the size of the transistor is all that matters – the smaller, the smarter. The day of *nanotechnology* has come to the fore and yesterday's gossip of 'atomic-sized' transistors has almost become today's reality. In the near future, transistors may measure less than 100 nanometers – compare this with the *vacuum tubes* of the first modern computer, the ENIAC, which measured about 3 inches and would have been able to accommodate about 1.5 billion-billion of these modern transistors.

THE NEURON AS THE CHIP

It is common practice to compare the functioning of the brain to a computer. In this section we will show that a single neuron can be compared to a computer – and the *molecular grid* is shown to have a remarkable resemblance to a *transistor*. First let us see the essential *differences* between computer and the brain, as summarized in Table 11.1.

TABLE 11.1 Differences between Computer and Brain

	Character	**Brain**	**Computer**
1	Signals	Electrochemical	Electrical
2	Number of fundamental units	More than 100 billion neurons	Merely hundreds of millions of transistors
3	Processing mode	Not yet known	Digital
4	Processing speed	Not fixed	Fixed
5	Memory recall	Not absolute, but unlimited	Relatively absolute, limited
6	Memory	Cannot be copied, pasted or deleted	Such tasks can be done
7	Capability	Learning and analysis better	Multitasking and speed better
8	Tasks	Innovative	Preprogrammed
9	Actions	Can be partial	All-or-none
10	Expandability	Innate plasticity	Rigid unless modified
11	Intellect	Yes	No
12	Commonsense	Yes	No
13	Working hours	24 h/day	Can be turned off
14	Imagination and creativity	Yes	No
15	Emotional interference	Very common	None
16	Will to survive	Present	Absent
17	How it works?	Not definitely known	Known, because man-made
18	Repair	Yes, occasionally	Needs regular overhaul
19	Cost	You can make a guess!	Costly
20	Future evolution?	Not in our hands	Yes, it is endless

TABLE 11.2 Summary of Comparison

	Category	Neuron	Computer
1	Electrical supply	Axon	Electrical cable
2	Transistors	Molecular grids	Metals
3	Chip	Dendritic pleat	Microprocessor
4	Short-term memory	Dendritic pleats with temporary molecular bonds	RAM
5	Long-term memory	Dendritic pleats with durable molecular bonds	Hard disk
6	Input devices	All sensory organs	Mouse, keyboard, etc.
7	Output devices	Muscles, blood vessels, glands	Monitor, sound system, etc.

A typical neuron consists of soma, axon and dendrites – all of which are electrically active and we have seen the functions of these structures in the preceding chapters. Now we will draw parallels between computer and neuron, as summarized in Table 11.2.

'Bio-electrical Supply'

Computer. A computer needs an electrical supply to work, which is supplied by an electrical cable
Neuron. For a neuron the supply comes from the axon of the preceding neuron in the form of action potentials (which in turn are produced initially by the peripheral environmental stimuli; see Ch. 3, p. 54).

'Biological Transistor'

Computer. A computer needs a triode (see above), which is made up of semiconductors sandwiched in three layers.
Neuron. Lipid membrane is an insulator (poor conductor of electricity) but when it is 'doped' with 'impurities' such as special lipids and water-soluble proteins, it becomes a semiconductor. Electrophysiologically, the molecular grids situated in the dendritic membrane of a neuron have three functional layers (see Ch. 7, p. 141 & Fig. 7.11) – a highly conducting extracellular fluid (ECF) and a highly conducting internal cytoplasm, and a semiconductor lipid layer sandwiched in between (Ch. 2, p. 36). These represent *triodes*, which can be called *'biological transistors'*.

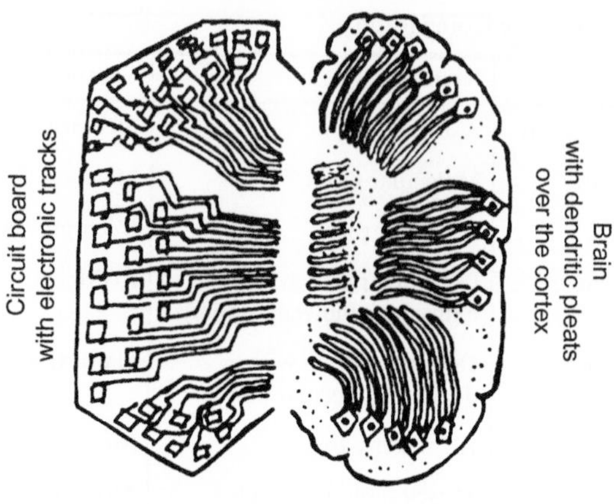

FIGURE 11.1 Microprocessor *versus* cerebral cortex.

'Biological Chips'

Computer. We have seen that there are millions of transistors in each microprocessor, all arranged in series.

Neuron. Perhaps the ***dendritic pleats*** represent the ***biological microprocessors*** (or *biological chips*) (Fig. 11.1) – each dendritic pleat containing variable number of molecular grids, all connected in series (Ch. 8, p. 155).

RAM and STM

Some analogy can be drawn between RAM and short-term memory: both of these tasks are done by sorting and storing temporary memory tasks – but the comparison should end there. Short-term memory is immensely innovative and intriguing – no machine can match it. Hard disks can be compared to long-term memory (Table 11.2).

Input/Output Devices

Computer. A computer has several input devices, such as mouse, keyboard, joystick, etc. The usual output devices of a computer are the monitor, sound system, etc.

Neuron. A neuron has numerous input devices in the form of sensory receptors in various sense organs such as eyes, ears, tongue, nose, etc. The neuron's output devices are muscles, blood vessels, glands and other tissues.

SOME HIGHLIGHTS IN THE COMPARISON

Processing Speed and Memory Capacity

It is estimated that the processing power of an average brain is about 100 million MIPS (million computer instructions per second). The memory capacity of the brain is really unlimited. There has been no news, to date, that someone is 'overloaded' with information and is required to 'shut down'. There are about 100 billion neurons in the human brain with more than 100 trillion synapses, and the human brain can theoretically hold millions of megabytes of memory!

Artificial Intelligence: Do Computers Surpass Human Brain?

To give them fair credit, computers are now routinely used to run large offices and factories. Present-day robots are being used to perform complex, delicate and monotonous tasks such as '*robotic surgeries*' and other delicate operations in some factories. But again, today's mantra is that *smaller is better*. If the electric circuit is smaller, the transmission *across* it is faster and the action is rapid. But the trick of smallest circuit has an apparent limit – the problem of 'electron leakage' and 'heat melting' may interfere with performance if size becomes *too* small. But now, the answer may lie in '*quantum computers*', which work on the principle of 'single electron transfer' and on 'quantum dots'. Today this seems to have become a reality with nanotechnology, especially that of carbon nanotubes.

However, there are certain dangers in such total dependence of human life on computers. The computers may be fast and accurate, but we have to weigh this benefit against the risk of people losing jobs, resulting in the devaluation of human skill and experience. Another problem with total dependence is that, although it may sound a far-fetched possibility, some electronic cataclysm (some electromagnetic flux, or an alien invasion?) may disrupt all digital mechanisms developed by us and wipe out all digital data – and leave us in utter 'digital darkness'! Another equally pessimistic possibility is that these digital machines may turn against their own masters, as shown in some sci-fi movies. But again, these are merely fantasies.

References

Anastassiou, C.A., Perin, R., Markram, H., Koch, C., 2011. Ephaptic coupling of cortical neurons. Nat. Neurosci. 14, 217–223.

Barrett, K.E., Barman, S.M., Boitano, S., Brooks, H.L. (Eds), 2012. Ganong's Review of Medical Physiology (twenty fourth edn). Tata McGraw Hill, New Delhi.

Borrie, R.A., 2002. The use of restricted environmental stimulation therapy in treating addictive behaviors. Int. J. Addict. 25 (7A–8A), 995–1015, 1990–1991.

Bruno, D., 2002. The Brain from Top to Bottom. Retrieved March 28, 2014 from website: <http://thebrain.mcgill.ca/>. Affiliation: Canadian Institute of Neurosciences, Mental Health and Addiction.

Durant, W., 2006. The Story of Philosophy. Simon & Schuster, New York.

Fine, C. (Ed.), 2008. Britannica Guide to the Brain. Encyclopedia Britannica Inc, Sunnyvale, CA.

Freund, T., Kali, S., 2008. Interneurons. In: Izhikevich, E.M. (Ed.), Scholarpedia 3 (9), 4720. Retrieved March 28, 2014 from website: <http://scholarpedia.org/>.

Guyton, A.C., Hall, J.E. (Eds), 2008. Textbook of Medical Physiology (eleventh edn). Elsevier-Saunders, Noida.

Han, J., Kesner, P., Metna-Laurent, M., et al. 2012. Acute cannabinoids impair working memory through astroglial CB_1 receptor modulation of hippocampal LTD. Cell 148, 1039–1050.

Hardin, J., Bertoni, G., Kleinsmith, L.J. (Eds), 2012. Becker's World of the Cell (eighth edn). Pearson Education, San Francisco (International edn).

Maxfield, F.R., 2002. Plasma membrane microdomains. Curr. Opin. Cell Biol. 14 (4), 483–487.

Noble, C.H., Rowland, C.F., Pine, J.M., 2011. Comprehension of argument structure and semantic roles: evidence from English-learning children and forced-choice pointing paradigm. Cogn. Sci. 35 (5), 963–982.

O'Connor, T., 2013. In: Zalta, E.N. (Ed.), The Stanford Encyclopedia of Philosophy. Retrieved March 28, 2014 from URL: <http://plato.stanford.edu/archives/spr/entries/freewill/>.

Ossandón, T., Jebri, K., Vidal, J.R., et al. 2011. Power in the default-mode network is correlated with task complexity and subject performance. J. Neurosci. 31 (41), 14521–14530.

Simons, K., Toomre, D., 2000. Lipid rafts and signal transduction. Nat. Rev. Mol. Cell Biol. 1 (1), 31–39.

Singh, M., 2005. Essential fatty acids, DHA and human brain. Indian J. Pediatr. 72 (3), 239–242.

Spira, F., Mueller, N.S., Beck, G., von Olshausen, P., Beig, J., Wedlich-Söldner, R., 2012. Patchwork organization of the yeast plasma membrane into numerous coexisting domains. Nat. Cell Biol. 14 (6), 640–648.

van den Bogaart, G., Meyenberg, K., Risselada, H.J., et al. 2011. Membrane protein sequestering by ionic protein-lipid interactions. Nature 479 (7374), 552–555.

Wang, W., Zhu, J.Z., Chang, K.T., Min, K.T., 2012. DSCR1 interacts with FMRP and is required for spine morphogenesis and local protein synthesis. EMBO J. 31 (18), 3655–3666.

Zelano, C., Mohanty, A., Gottfried, J.A., 2011. Olfactory predictive codes and stimulus templates in piriform cortex. Neuron 72, 178–187.

Glossary

A list of specific words or phrases used in the book is given below. Their related words are given in brackets for a quick cross-reference.

A posteriori memory Memory derived from experience. An idea, which is an assembly of primary thoughts, is an a posteriori phenomenon. (*A priori memory.*)

A priori memory An innate memory that exist prior to experience. All primary thoughts are *a priori* phenomena. (*A posteriori memory.*)

Action potentials Action potentials are large electrical potentials generated by the neuronal membranes upon their excitation. They usually have a typical voltage potential of about 100 mV, and are characteristically transmitted by the axons. (*Graded potentials.*)

Awareness Awareness is the process of sensing the external environment, and it is the characteristic feature of all life forms on earth. (*Consciousness, mind.*)

Axodendritic synapses Axons synapse with the dendrites, thus the signals pass from axon to dendrite. This is the commonest form of communication between neurons in the cortex.

Barcode arrangement A particular arrangement of proteins and lipids in a molecular grid which confers it specificity. This specific arrangement results in the generation of specific thought-potentials that are converted into characteristic primary thoughts.

Biological transistors Three layers of semiconductors form a transistor in a computer – likewise, three layers of 'biological semiconductors' in the cell membrane make a biological transistor – the three layers being the extracellular fluid, cytoplasm and membrane. Biological transistors are fundamentally responsible for the electrical activity of the neuronal membrane.

Central executive The superb but ethereal agent of the human mind, which is responsible for organizing primary thoughts and ideas in order to give us a meaningful idea of the external world. The central executive is unique to each individual.

Chemical synapse The commonest mode of communication between the neurons in the CNS. They use neurotransmitters as mediators. (*Electrical synapse.*)

Complex ideas A complex form of idea containing a number of primary thoughts, usually collected from different cortical areas. (*Simple idea.*)

Consciousness An advanced form of awareness executed by the nervous system, and it is the characteristic feature of all animals. Consciousness is responsible for memory. (*Awareness, mind.*)

Dendritic plasticity An important type of synaptic plasticity where the dendritic spines of a dendritic tree may appear or disappear in order to make or break synaptic connections with axons. It is a rapidly growing concept in neuroscience. (*Synaptic plasticity.*)

Dendritic pleats Many dendrites containing molecular grids arranged together in pleats, causing them to excite together to produce an array of primary thoughts simultaneously. These arrays of primary thoughts form ideas.

Dendritic spines These are membranous protrusions on the dendrites that chiefly participate in synapse formation and are responsible for dendritic plasticity. (*Dendritic plasticity.*)

Dendrodendritic synapses Neurons communicating quickly between themselves via their dendrites, thus bypassing the relatively slow axodendritic synapses.

Determinism The philosophical doctrine that external world events are determined and are independent of human interference, and thus are invariant. (*Indeterminism.*)

Electrical synapse A faster mode of communication between the neurons (or other excitable tissues). They are mediated by gap junctions, and do not use neurotransmitters. (*Chemical synapse.*)

Free will Free will is the capability of a rational agent (human being) in executing freedom in acts of conscious choice without the constraint of necessity or fate.

Graded potentials Graded potentials are small electrical potentials which vary in their amplitude between 1 and 50 mV. They are usually produced by the action of neurotransmitters on the dendrites. (*Action potentials.*)

Idealism A philosophical doctrine which presumes that a person's thoughts alone are the fundamental reality and the external physical world is only a reflection of our ideas, and there is no way by which we can know the real nature of things. (*Realism.*)

Ideas An assembly of primary thoughts.

Indeterminism The philosophical doctrine that the events of the external world are dependent on human or any other interference, and thus are variable. (*Determinism.*)

Labeled-line principle Even though nerves conduct only uniform impulses to the brain, humans experience different modalities of sensations, because the signals reach specific areas of the brain.

Lipid raft Lipid rafts are localized thickenings in the cell membrane containing an aggregation of various kinds of lipids. They are the fundamental constituents of molecular grids.

Memory traces The biophysical and biochemical changes that occur in neuronal networks when memory is stored. It is the hardware of memory.

Microdomains Microdomains are specialized lipid rafts in the cell membrane, which incorporate a variety of proteins (especially the transmembrane proteins) and become functional units of cell membrane. Many microdomains together form molecular grids.

Mind The advanced form of consciousness manifested in human beings, perhaps also in some higher animals. (*Awareness, consciousness.*)

Molecular grid A molecular gadget located in the neuronal membrane, which is responsible for converting an external stimulus into a primary thought. Molecular grids are composed of a group of microdomains. Upon excitation they generate primary thoughts.

Perceptual neurons Perceptual neurons are special sensory neurons present in the CNS that can convert external stimuli into primary thoughts.

Phosphenes The phenomenon of perception of sparkles of light experienced by us when the retina is stimulated by energies other than light.

Primary thoughts Primary thoughts are the fundamental units of thought occurring at the level of neurons. They are generated by the molecular grids located in the neurons by converting external stimuli into internal perceptions. (*Ideas.*)

Realism A philosophical doctrine that presumes the external world exists independently of us and the events of the external world go on and on, irrespective of the observer's status. (*Idealism.*)

Simple ideas An assembly of a limited number of primary thoughts formed usually at a single cortical area. (*Complex ideas.*)

Synaptic delay The time taken for the signals to pass across a synapse. It is maximum for a chemical synapse and minimum for an electrical synapse.

Synaptic plasticity Synaptic plasticity is the ability of neurons to bring about changes in the connections between neuronal networks in response to use or disuse. It is the most important property of nervous system which facilitates memory and learning. (*Dendritic plasticity.*)

Thought-potentials The action potentials generated at the dendrites, which are finally converted into primary thoughts. They are generated by voltage ion channels located in the dendritic membrane.

Volition Volition is the ability of an animal to demonstrate any action which is *not* directly related to the will to survive. (*Will to survive.*)

Whole neuron concept The simplistic notion that the neuron as a whole is responsible for the generation of thought, whereas most current concepts in science, if not all, are reductionist in their approach.

Will to survive (or simply, the will) This is the basic urge of all organisms to live and it is the fundamental property of life, which is manifested perpetually in every organism. (*Volition.*)

Index

Note: Page numbers followed by *"f"* and *"t"* refer to figures and tables respectively.

D

M

N

O

P

Q

R

S

T

U

Printed and bound by CPI Group (UK) Ltd, Croydon, CR0 4YY

03/10/2024

01040425-0003